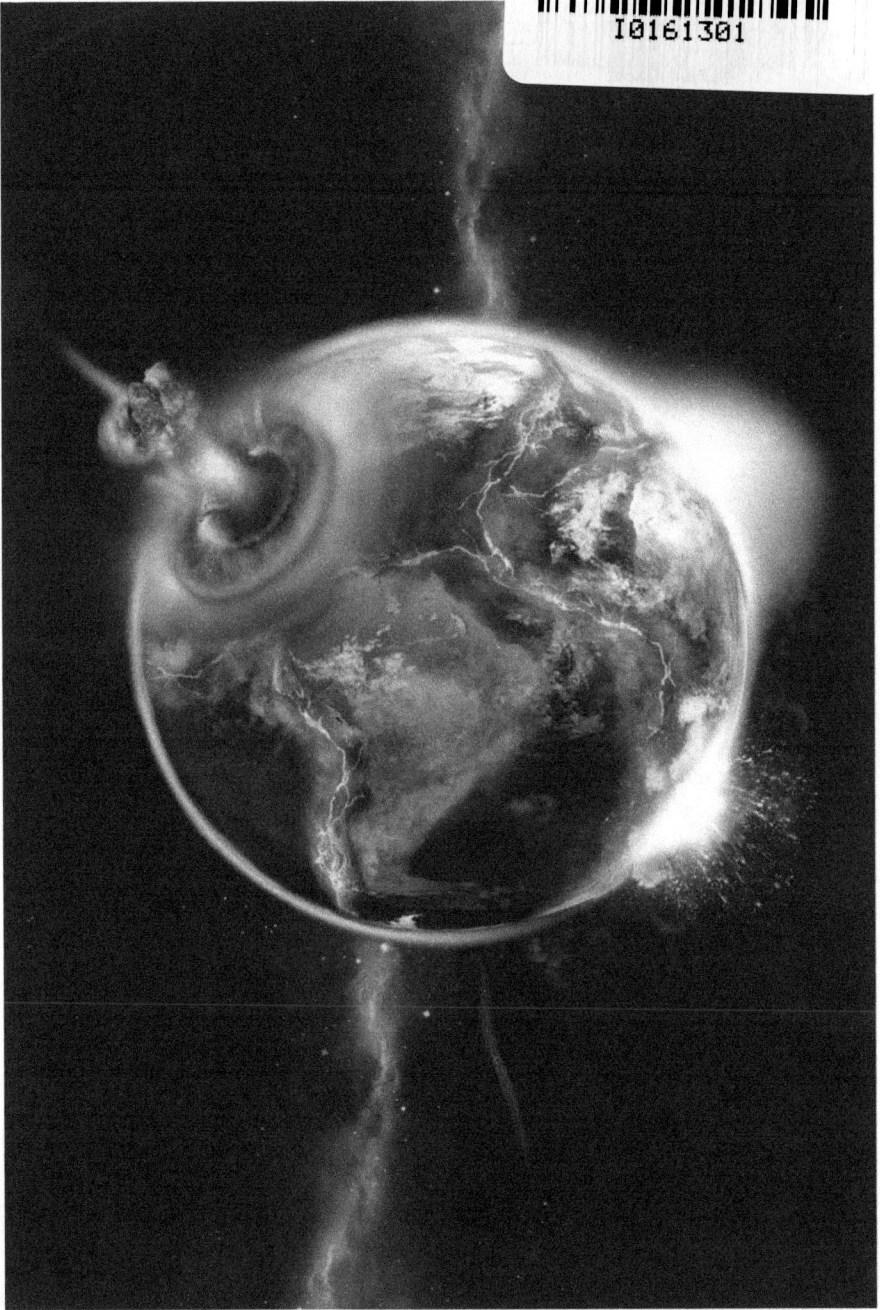

D.H. ALEXANDER

Cover art by Jose Razo.
Illustrations by Loren Rodriguez.

The Front Cover illustrates the consequences of a large impactor. Notice that opposite the point of impact, an outburst of volcanism is taking place. Scientists believe that the Chicxulub impact wiped out the dinosaurs. It was also responsible for the catastrophic outpouring of basalts on the Deccan Plateau antipodal to the impact (at a point on the globe opposite the impact). The impact also triggered basalt magma outpourings on the ocean floors. In the years that followed the impact, representing about 1.5% of geologic time, nearly 65% of new ocean floors were deposited as illustrated on the back cover.

Accounts based upon actual events.

ISBN: 978-1-952229-02-2

Sword Bearers Ministries
PO Box 2010 Richland, Washington

Preface

Invisible things of him from the creation of the world are clearly seen, being understood by the things that are made, [even] his eternal power and Godhead; so that they are without excuse...[1] If thou seekest her as silver, and searchest for her as [for] hid treasures; Then shalt thou understand ...[2] For the stone shall cry out [3]...

Collective libraries of the ancient and modern worlds store an incredible amount of knowledge. You might say, the rocks of the Earth provide a complementary library for the trained scientist.   Each **Rock Volume** of our **Earth's Chronicles,** documents numerous asteroid impacts, extensive volcanic outpourings, extinctions, rotation of the Earth's plates, polar wobble, magnetic field fluctuation, flooding, and more. Synthesis of Phanerozoic events reveal regularly repeated patterns of past Apocalyptic cycles.   Importantly, assembly of evidence from *Nature's Library* provides a means of probing the planet's future. **_And scientists are finding that the Rocks are Crying Out, warning mankind of a 7th APOCALYPSE NOW set in motion._**

A new generation of post-Einsteinian geologists, geophysicists, geochemists, paleontologists, biologists, physicists, cosmologists and a myriad of other scientists, like myself, arrived on the scene late in the last century armed with incredible tools, not available to our predecessors a mere generation earlier. Our new generation would forever change man's understanding of the living, breathing planet that we call home. What we learned would challenge the very foundations and even topple the **"*hypotheses*"** and **"*theories*"** of generations past. These tools, coupled with the time-tested scientific methods of our predecessors, would enable us to unlock hidden secrets of the Earth, planets, and moons of our Solar

---

1 Romans 1:20
2 Proverbs 2:3-6.
3 Habakkuk 2:11.

System; even the Universe. Armed with deep space probes, satellite telescopes, and new technologies we have remotely scanned the Universe and Earth's land surface, revealing a living Planet never understood to such depths before. We've been able to probe the compositions of distant galaxies in an insatiable quest for the knowledge of the origin of the Universe and life itself. But the secrets of the sea floor, amounting to some 70% of Earth's surface, remain largely locked to this day. Some of us have probed the Earth's interior, confirming that we live on a planet shattered by asteroids and driven by **Plate Tectonics**. Seismologists, armed with seismic tomography and heat flow data, image the Earth's interior as a physician uses an ultrasound to provide a sonogram of a fetus in the womb. Some have accumulated paleomagnetic data that confirm that the Earth's surface had a violent past of colliding plates driven by such incredible forces that they thrust deep sea floors miles upward, whence they became mountain peaks! Others have assembled deep borehole information, from core samples, electric well-logging, and other deep hole techniques like nuclear magnetic resonance, to assemble a picture of the Earth's upper crust just as physicians use it to produce MRI's. Attempts to drill to the transition layer between the Earth's crust and mantle (Mohorovičić discontinuity) still remain unsuccessful. Yet we know that the Earth's crust is less than ~1% of the thickness of its interior. You might say **we've only scratched the surface** of Earth.

Since the **sudden appearance of life,** in just the last 12% of Earth's geologic history, Earth has been through one global catastrophe after another, including 6 cycles of global flooding and 6 devastating extinctions. During these cycles, continents have been formed and destroyed resulting in a complete resurfacing of the Earth. And in just the last ~1.5% of **Earth's Chronicles** vast seafloors beneath the oceans, even the Pacific, have been largely replaced by the rapid invasion of noxious gases and magma now turned to rock. Are these cycles an interwoven continuum? **Based on the scientific evidence, a 7th Apocalypse is even now in its formative stages. Alarmingly, this 7th Apocalyptic Cycle seems to be leading to a catastrophic, and potentially fiery end to the planet, leaving mankind's existence weighed in the balance.**

Dedication

In a watershed moment in time, the worlds of science and religion were turned upside down. In 1733, Sir Isaac Newton's book on prophecy, ***Observations upon the Prophecies of Daniel, and the Apocalypse of St. John***[4] was published. A hundred years later, Charles Lyell completed the publishing of his three volume set, ***Principles of Geology***,[5] which applied the ***concept of uniformitarianism***. This concept provides the foundation for Darwin's ***Origin of the Species***.[6] And by doing so, ***Lyell helped push the world into the Last Days of Earth's history*** by triggering the fulfillment of the 5 prophecies of 2nd Peter 3:3-6. Newton stood on one side of the chasm where science, the Scriptures, and God were once united. Lyell and Darwin stood on the other side jettisoning, and thus separating *"science"* from its reliance upon the Scriptures and upon God. How could the apostle-writer have known that [1] scoffers would [2] denounce Christ's Second Coming, [3] embrace uniformitarianism and evolution, [4] deny a Creator God, and [5] deny Noah's Flood? ***This book presents the scientific basis for the catastrophic origin of the world and challenges uniformitarianism and evolution.***

This book is dedicated to my colleagues in the realm of science, my students, my wife, my parents, and those that I have adopted as my brothers and sisters, my sons, my daughters, my grandchildren, my friends and co-workers, and my illustrators and editors.

---

4 Newton, Isaac, Sir. Observations upon the Prophecies of Daniel, and the Apocalypse of St. John. London. J. Darby and T. Browne. 1733.
5 Lyell, Charles. Principles of Geology. 1st Edition. John Murray, London. Volume 1, January 1830; Volume 2, January 1832; Volume 3, May 1833.
6 Darwin, Charles, 1809-1882. On The Origin of Species by Means of Natural Selection, or Preservation of Favoured Races in the Struggle for Life. London :John Murray, 1859.

Table of Contents

He that sits above the wheels of the heavenly chariot still guides the affairs of mankind[7]

In the context of the vastness of space and the immensity of time we are as nothing. Why would anyone care?

Soon there will be Time no Longer![8]
Is God in control of your destiny?

---

7 Ezekiel 10:1-4; Ezekiel 1:4-28.
8 Revelation 10:6.

CATASTROPHISM?

Behold, the LORD lays the earth waste, devastates it, distorts its surface and scatters its inhabitants.[9] *And all flesh died that moved upon the earth, both of fowl, and of cattle, and of beast, and of every creeping thing that creepeth upon the earth, and every man: All in whose nostrils [was] the breath of life, of all that [was] in the dry [land], died. And every living substance was destroyed which was upon the face of the ground, both man, and cattle, and the creeping things, and the fowl of the heaven; and they were destroyed from the earth...*[10]

Earth's history is punctuated by one catastrophe after another. Singular catastrophies weakly defy the concept of Uniformitarianism that lies at the very core of Evolution. Yet many authors are now proposing links between two or more catastrophic events, but is there an Apocalyptic pattern of events that shatters Evolution's foundation? We begin our journey by examining accepted catastrophic events, searching for clues to see **who took the right path. Newton believed Earth was Catastrophically Flooded. Darwin and Lyell changed the world of science to a belief in Uniformitarianism without Global Flooding. Who is right? You be the judge!**

---

9 Isaiah 24:1.
10 Genesis 7:21-23.

1 *Elements of Catastrophe: the Calm Before the Storm*

This matter [is] by the decree of the watchers, and the demand by the word of the holy ones: to the intent that the living may know that the most High ruleth in the kingdom of men...[11] [Seek him] that maketh the seven stars and Orion, and turneth the shadow of death into the morning, and maketh the day dark with night: <u>that calleth for the waters of the sea, and poureth them out upon the face of the earth</u>: The LORD [is] his name[12]

07:58AM 26 December 2004, Coast of Sri Lanka

Flamingos abandoned their nests and elephants screamed and raced for higher ground,[13] sensing the incoming tsunami tidal wave originating from the ruptured fault off the coast of Sumatra that would soon *pour floodwaters upon the face of the earth*. A fault of more than 900 miles long ripped open the ocean floor and sent waves to the coasts of Sri Lanka, India, and elsewhere. Terror filled the hearts of men as a series of tsunamis reaching 100 feet high raced inland by the rupture between the Burma and Indian tectonic plates. According to a BBC report:[14] *About 228,000 people were killed as a result of the <u>9.1 magnitude quake</u> and it's tsunamis that slammed into coastlines on 26 December 2004. The violent upward thrust of the ocean floor at 07:58 local time (00:58 GMT) displaced billions of tonnes of seawater, which then raced towards shorelines at*

---

11 Daniel 4:17.
12 Amos 5:8.
13 National Geographic. M. Mott. January 4, 2005. Did Animals Sense Tsunami was Coming?
14 BBC News, December 25, 2014. Indian Ocean Tsunami: Then and Now.

*terrifying speeds. **The planet vibrated as much as a half an inch,**[15] **and triggered earthquakes as far away as Alaska.**[16] What kind of forces lurk below the surface of the planet that could cause such damage? **What vibration would a larger fault rupture have?***

10:00AM Monday, 27 August 1883 Sunda Strait, Dutch East Indies

The sailors reacted in terror as the loud explosion damaged their ear drums and was closely followed by hurricane force winds that nearly capsized their ship. A tower of ash and hot rock shot into the sky, as the paradise island that once was at the Sunda Strait between Sumatra and Java was no more. With the mighty explosion, the island of Krakatoa collapsed beneath the ocean surface. Powerful shock waves were registered as they travelled around the globe nearly 4 times. The blast was heard in Perth Australia, nearly 2000 miles away. In the end, over 36,000 were killed by the explosion and the tsunamis that raced upon land. Nearly 135 years later, the Indonesian town of Merak was destroyed by a tsunami nearly 140 feet high.[17] On December 22, 2018, the eruption and collapse of Anak Krakatoa (referred to as the *Child of Krakatoa*) resulted in over 400 deaths and 14,000 injuries.[18] The earthquakes and volcanic eruptions near Sumatra and Java are related to the large 900 mile long "crack" in the earth where two giant pieces of the planet are colliding with one another. Along this plate boundary, the explosions of Krakatoa (1883) and Tambora (1815) released such massive amounts of ash and sulfur gases that they caused global cooling. The ash from **Tambora** lowered global temperatures, in an event sometimes known as

---

15 Walton, Marsha (20 May 2005). "Scientists: Sumatra quake longest ever recorded". *CNN.*

16 West, Michael; Sanches, John J.; McNutt, Stephen R. (20 May 2005). *"Periodically Triggered Seismicity at Mount Wrangell, Alaska, After the Sumatra Earthquake". Science.* **308** (5725): 1144–1146.

17 Winchester, Simon. 2003, Krakatoa: *The Day the World Exploded: August 27, 1883.*

18 The Star Online. December 31, 2018. Number of Injured in Indonesia Tsunami surges to over 14,000.

the Year Without a Summer in 1816.[19] Historical data from the Pacific Ocean plate boundaries (known as *the Ring of Fire*) show nearly 150,000 <u>known</u> deaths![20] **If these volcanoes caused global cooling, could numerous volcanic eruptions cause glaciation?[21]**

11:49AM September 8, 2017 Chiapas, Mexico

On September 8, 2017 a large 8.2 magnitude earthquake hit Chiapas, Mexico near the border with Guatemala. The quake and 12 aftershocks shook Mexico City.

> *"We don't yet have an explanation on how this was possible," said [Diego] Melgar, an assistant professor in the <u>Department of Earth Sciences</u>, [University of Oregon]. "We can only say that it contradicts the models that we have so far and indicates that we have to do more work to understand it." Subduction zone megaquakes generally occur near the top of where plates converge. Initially, the 2017 event was thought to be in such a location, where the Cocos ocean plate is being overridden, or subducted, by a continental plate. The zone had not had a quake of such magnitude since 1787. The epicenter, however, was 28 miles deep in the Cocos plate, well under the overriding plate and where modeling had said a quake shouldn't happen.[22]*

Powerful earthquakes occur where one plate is pushing and eventually sliding (or subducting) beneath or over another. Tectonic activity along plate boundaries cause earthquakes, tsunamis, destructive volcanic eruptions, and turbidites (underwater mud flows). Analysts now say that the 2017 event was triggered by the subduction of the Cocos plate under the North

---

19 Kious and Tilling, 1996. *This Dynamic Earth: The Story of Plate Tectonics:* USGS General Interest Publication.

20 1815 Eruption of Tambora. Wikipedia.

21 University of Exeter. 2015. *Volcanic event caused ice age during Jurassic Period.* ScienceDaily.

22 University of Oregon. October 25, 2018. *2017 Mexico quake came from unexpected location, study says.*

American plate. The breakage of the Cocos plate affected more than 1.5 million people and damaged over 40,000 homes.[23]

9:20AM February 15, 2013 Chelyabinsk, Russia

The BBC News reported that Sergei Serskov was working in his office at about 9:20 in the morning of February 15, 2013 in the city of Chelyabinsk in the Russian Urals when a bright streak of light passed through the window. *"I looked out the window and saw a huge line of smoke, like you get from a plane but many times bigger. A few minutes later the window suddenly came open and there was a huge explosion, followed by lots of little explosions. It felt like a war zone and it lasted about 20 to 30 minutes."[24]* School children raced to their classroom windows to watch the spectacle. According to news reports, some thought aliens were attacking, others thought the sun was falling. The 20 meter bolide exploded at an altitude of about 12 miles above Earth while traveling at nearly 40,000 miles per hour, releasing 500 kilotons of energy, approximately *30 times the yield of the nuclear bomb over Hiroshima*. It caused a shock wave that broke windows in six Russian cities and caused about 1,500 people to seek treatment for injuries, mainly from flying glass. The explosion glowed 30 times brighter than the sun and observers had to look away because it was so bright it literally sunburned observers. People near Chelyabinsk, Russia felt the ground shake, smelled the sour stench of sulfur, and heard windows shatter. In Chelyabinsk alone, more than 3,500 buildings were damaged. Researchers found shockwave destruction as far as 100 kilometers away from the impact site.

An estimated *25 million meteoroids, micrometeoroids, and other space debris enter Earth's atmosphere each day*.[25] Scientists believe that the Earth has suffered numerous major impacts, each capable of destroying much of the life on the planet.

---

23 Ramirez-Herrera, Maria, et al. Pure and Applied Geophysics January 2018, Volume 175, Issue 1, pp 25–34. The 8 September 2017 Tsunami Triggered by the *M*w 8.2 Intraplate Earthquake, Chiapas, Mexico
24 BBC News, February 15, 2013. *Russia meteor eyewitness: 'Something like the sun fell.'*
25 Kidz, Franz, January 2019. The Oldest Material in the Smithsonian Institution Came from Outer Space. Smithsonian Magazine. Smithsonian.com

But could these same objects that bring death also be the source of the seeds of life?

1991 Mission to Mayak, Siberia, Former Soviet Union

I took special interest in the event because I had led a team of scientists to the nearby nuclear city of Mayak two decades earlier. Just after the East Berlin wall fell, I was privileged to lead a team of scientists from US DOE National Laboratories to Chelyabinsk, Siberia, Russia. It is the site of a 1957 nuclear waste tank explosion that spread radionuclides over a vast farming region. Our team advised our hosts of ways of containing and cleaning up radioactive wastes; expressly deep groundwater wastes that could be treated by pumping and filtering out radwastes. More than 250,000 cases of cancer had been caused by the radioactive waste along the Techya River. So we negotiated for cancer data that the Russians had collected on the regional population because we had no such data in the U.S. We also negotiated the exchange of U.S. and Russian scientists to collaborate in the study of nuclear waste.

Fall 1974, Sudbury, Ontario

In the Fall of 1974, I accompanied a team of fellow graduate students led by Dr. Eric Essene, an accomplished metamorphic petrologist, to the Sudbury impact site in Ontario, Canada. Meteoritic impact sites leave a signature that is unmistakable to the trained eye. Everywhere we looked, we found shatter cones, shattered minerals, and metamorphosed features that resulted from the impact. Of the discovered impact sites to date, the Sudbury bolide is second only to the one that created the 190 mile wide crater at Vredefort, Africa. The Sudbury crater is about 250 kilometers in diameter and its bolide is estimated to be on the order of 14,000 meters (14 kilometers) which dwarfs the 20 meter bolide that raced across the sky at Chelyabinsk, Russia.[26] Similarly, the famous Chicxulub crater under the Yucatan peninsula of Mexico is credited by some as the cause of the extinction of the dinosaurs[27]

---

26 Deutsch, A. 1994. *Isotope Systematics Support the Impact Origin of the Sudbury.*
27 Alvarez, L.W.; Alvarez, W.; Asaro, F.; Michel, H. V. 1980. "Extraterrestrial cause for the Cretaceous–Tertiary extinction". *Science.* **208** (4448): 1095–1108.

and is about 93 miles in diameter. If the 20 meter diameter bolide released the energy equivalent of approximately 30 times the yield of the nuclear bomb over Hiroshima, consider what the bolides at Sudbury, Vredefort, and Chicxulub must have done to the planet! The collision at Chicxulub is estimated at 100,000,000 megatons of TNT $(4.2\times10^{23}$ J), **over a billion times the energy of the atomic bombs dropped on Hiroshima and Nagasaki.**[28] And some think the bolide that created the Sudbury crater could be the largest of the three.[29] *Debris from the [Sudbury] impact was scattered over an area of 1,600,000 km² (620,000 sq mi) ... rock fragments ejected by the impact have been found as far as Minnesota.*[30] **Could a series of combined impacts have moved mountains or triggered the breakup of the Earth's giant plates? Could they have affected the Earth's tilt? I think so!**

4:53 PM January 12, 2010, Port-Au Prince, Haiti

Deep within the Earth, pressures were building along the Enriquillo-Plantain Garden Fault along the margins of two of the Planet's tectonic plates. According to the US Geological Survey, these plates can be traced to the Mesozoic breakup of Pangea and the creation of the Atlantic Ocean.

On 12 January 2010, at 4:53 p.m. local time, a magnitude 7.0 earthquake struck the Republic of Haiti, with an epicenter located approximately 25 km south and west of the capital city of Port-au-Prince. Near the epicenter of the earthquake, in the city of Leogane, it is estimated that 80%– 90% of the buildings were critically damaged or destroyed. The metropolitan Port-au-Prince region, which includes the cities of Carrefour, Pe'tion-Ville, Delmas, Tabarre, Cite Soleil, and Kenscoff, was also severely affected. According to the Government of Haiti, the earthquake left more than 316,000 dead or missing, 300,0001 injured, and over 1.3

28 Schulte, P.; et al. (2010). "The Chicxulub Asteroid Impact and Mass Extinction at the Cretaceous-Paleogene Boundary". *Science*. 1214–1218.
29 "The Vredefort Dome: Centre of the World's Largest Meteorite Impact Structure!" (source: Wikipedia)
30 Associated Press: "Ontario crater debris found in Minn.", *Star Tribune*, July 15, 2007, (source: W8ikipedia).

million homeless (GOH 2010). According to the Inter-American Development Bank (IDB) the earthquake was <u>the most destructive event any country has experienced in modern times when measured in terms of the number of people killed as a percentage of the country's population</u> (Cavallo et al. 2010).[31]

According to reports, seismic activity had been concentrated in the Dominican side of the fault but little pressure had been released in Haiti. Consequently, pressures had been building until the fault released an enormous amount of energy. Scientists fear a similar buildup along the San Andreas Fault could release considerable energy with severe consequences for San Francisco.

5:02 PM November 18, 1929 Newfoundland

On November 18, 1929 a turbidity current carrying an enormous quantity of rock fragments, mud, and exceedingly fine grained silt raced across the Atlantic Ocean, cutting 12 trans-Atlantic cables on its way to Portugal. The turbidite was triggered by a magnitude 7.2 earthquake off the coast of Newfoundland, Canada. The landslide triggered by the earthquake, not only produced the massive turbidite mudflows but also triggered a series of destructive tsunami waves that killed 28 people on the Burin Peninsula. *The turbidity current on the Grand Banks reached estimated speeds of 60 to 100 kilometers per hour and carried an astounding 100 to 150 cubic kilometers of mud and sand. The massive volume of displaced sediment in the landslide also caused a tsunami, which radiated more than 4,000 kilometers across the Atlantic, making landfall in Portugal more than six hours after the quake.[32]*

Imagine the magnitude of turbidites triggered by continental collisions! Turbidites are suspensions of mud like clouds that are often associated with the sudden burial of the famous soft-bodied fossils of the Cambrian Explosion.

---

31 DesRoches, R. and others. 2011. Overview of the 2010 Haiti Earthquake. Earthquake Spectra, Volume 27, No. S1, pages S1–S21.
32 Derouin, Sarah. November 18, 2017. Benchmarks: November 18, 1929: Turbidity currents snap trans-Atlantic cables. Earthmagazine.org.

July 1970, Glacier National Park, Montana

I had witnessed first-hand the power of nature in my early years as a geologist. At Glacier National Park, I viewed the *"Lewis and Clark Overthrust."* Geologists estimate that a slab of rock 3 miles thick and over 60 miles in length slid some 35 miles in response to an uplift to the west of the region caused by plate tectonics. Older rock layers that can be seen in Chief Mountain were thrust up over younger rock. What kind of forces are at work beneath the Earth's shattered plates that can thrust enormous slabs of rocks, the size of mountains, on top of younger rock? What forces would it take to raise enormous slabs of rock from the sea floor and elevate them to the peaks of mountains? What conditions allowed such an enormous slab of rock to be transported such a distance without significant damage? The collision of the Indian plate with the Eurasian plate led to the formation of the Himalayan Mountains.[33] Marine fossils and shells found in the limestones beds are geologic evidence of the ancient Tethys sea floor. Imagine ancient sea floors being forced upwards into mountain ranges several miles above sea level! **How could these mountain ranges be the product of uniformitarianism?**

Summer, 1922 Columbia Basin, Eastern Washington

It was a hot summer day in 1922. The geologist stood by the side of the road staring at the canyon-like coulee landscape wearing a distinctive construction worker's metallic helmet and smoking a pipe. Like many an Associate Professor,[34] he spent his summers conducting field work. But this was an exceptional summer. He was wrapping up his new theory that would transform the field of Geology. In 1923, geologist J. Harlan Bretz published observations that the basalt flows near the Grand Coulee in Washington State were deeply eroded in forms he called the channeled scablands,[35] concluding that the erosion had to be due to a large catastrophic flood. His findings set off a firestorm in the

---

33 USGS. 2015. The Himalayas: Two Continents Collide.
34 Tate, Cassandra. November 29, 2007. Bretz, J. Harlen (1882-1981), Geologist. HistoryLink.org
35 Bretz, J Harlen (1923). "The Channeled Scabland of the Columbia Plateau". *Journal of Geology.* 31 (8): 617–649.

scientific community because his conclusion contradicted the widely accepted model of uniformitarianism set forth by Charles Lyell. In 1925, Bretz issued a second paper,[36] once again concluding that the Columbia Basin had been the site of an enormous catastrophic flood. For the next forty years, Bretz was at the epicenter of controversy within the geological sciences. In 1927, J. Harlan Bretz was invited to a meeting hosted by the Geologic Society of Washington, D.C. In addition to Bretz, at least 6 other geologists presented competing theories on the origins of the Columbia Basin landscape. The discussions were heated and continued for decades after the meeting. Joseph Pardee, a geologist with the U.S. Geological Survey, attended the meeting and had serendipitously been studying the strand lines and other evidence of a large ancient lake in Missoula, Montana. He realized that ancient lake Missoula could well be the flood water source for the features that Bretz observed in eastern Washington State. Bretz and Pardee would team up to study the basin for nearly 30 years. The *"heretical"* theory proposed by Bretz was hotly contested, not only because it challenged the theory of uniformitarianism, but because it appeared to support the **Biblical Flood** and was called an **outrageous hypothesis**.[37]

But in 1972, NASA and the Geological Survey released satellite imagery of the Columbia Basin vindicating the theory put forward by Bretz. What appeared to take millions of years of erosion took place in days. Today scientists estimate that the Missoula flood height rose to nearly 1250 feet over Pasco, Washington, 1000 feet at the Dalles, Oregon about 315 miles away, and 400 feet at Portland, Oregon nearly 400 miles away.[38] *In only a few hours the megaflood cut canyons, potholes, waterfalls, and carried rock and sediment nearly a 1000 miles from the source.* The Missoula Flood that eroded and carved much of the Columbia Basin crested at the Wallula Gap between the borders of Washington and Oregon

---

36 Bretz, J Harlen (1925). "The Spokane flood beyond the Channeled Scablands". *Journal of Geology*. 33 (2): 97–115, 236–259.

37 Cassandra Tate, Bretzm J Harlen (1882-1981), Geologist. HistoryLink.org Essay 8382. Posted November 27, 2007.

38 The Columbia River, A Photographic Journey. *Missoula Floods*. Columbiariverimages.com/regions/places/Missoula_floods.html.

to a depth of about 1250 feet. It travelled at about 65 mph and left mountains of gravel 30 stories high, scattered 200 ton boulders like pebbles, and left giant ripple marks nearly 30 feet high.

I remember viewing the Camas Prairie ripple marks found along Highway 382 in Sanders County, Montana. It was a shock and awe moment for me as I realized the magnitude and power of the rushing torrent. Rock ripples some 35 feet high were left in its wake. And the wavelengths of the distinctive ripples were about a hundred feet apart; a far cry from the sand ripples you see on ocean beaches. Deposits of fine grained silts are found in the fringes of the Missoula Flood in places like the Willamette Valley of Oregon and near Walla Walla, Washington where the flood water velocities slowed. The fingerprint of mega floods are being found all around the planet. These include floods that carved the English Channel, separating England from Europe[39] and even the catastrophic refilling of the Mediterranean Sea. Catastrophes are clearly recorded in the Rock Record throughout the sedimentary layers of the Phanerozoic since the Cambrian Great Unconformity.

3:00 July 1984 Maotian Hill, Yunnan Province, China

Professor Hou Xian-guang completed the final leg of his journey in a hand drawn cart to join with the team from the Geological Bureau of Yunnan Province. [40] The team would look for a complete section of the Cambrian fossil beds where they had found the bradoriid arthropods in the past. On July 1st, the team was destined to make one of the greatest discoveries in the history of paleontology at Maotian Hill. As was his routine, Professor Xian-guang split open a mudstone in search of the usual trilobites and fossils. As he looked at the broken stone in his hand, he unexpectedly saw the white semi-circular film of an unknown fossil species. In his excitement, he and his team began opening heaps of other samples. They would soon find dozens of new biological species. They would also realize that they were working on a gold mine of **Earth's Chronicles**. The Chinese locality has

---

39 Schiermeier, Quirin 2007. The megaflood that made Britain an island. *Nature.*
40 Hou Xianguang, et al. April 2008. The Cambrian Fossils of Chengjiang, China: The Flowering of Early Animal Life. John Wiley. 248 pages.

been compared to the Burgess Shale deposits of Canada where the Cambrian explosion of life was first discovered by Charles Wolcott in 1909. In the days and years that followed, the Cheng-jiang Cambrian fossil discoveries would become world renowned. The soft bodied fossils are so well preserved that it is thought the fossil assemblage was likely buried by *periodic **turbidity currents**, since most fossils do not show evidence of post-mortem transport.*[41] But most importantly, *the fossil assemblages at Maotian Hill and in the Burgess Shale deposits defy the theory of evolution because there are no apparent predecessor life forms*. The same is true for the preservation conditions of the fossils in the Burgess shale.[42,43]

Most of today's ~36 living phyla are represented in these deposits including the Chordata phylum that includes animals even mammals, including humans. The diversity of Cambrian fossils and the inexplicable *lack of clear evolutionary precursor fossils* was known to Darwin and he saw it as a threat to his theory of evolution through natural selection. Stephen C. Meyers book, *Darwin's Doubt: The Explosive Origin of Animal Life and the Case for Intelligent Design*[44] makes the case that the fossil record points to an Intelligent Designer. You be the judge.

1972 Oklo, Gabon: Natural Nuclear Reactor Points to other deep-seated Natural Reactors

In 1956, Geochemistry Professor Dr. Paul Kazuo Kuroda,[45] mentor of two of my close professional friends, predicted that conditions favorable for a nuclear reactor could exist in the Earth. In 1972, French physicist, Francis Perrin discovered just such a deposit in

---

41 Maotianshan Shales. Wikipedia.
42 Gaines, Robert, R. 2014. *Burgess Shale-Type Preservation and it's Distribution in Space and Time*. Geology Department, Pomona College.
43 Petrovich, Radomir. 2001. *Mechanisms of Fossilization of Soft-Bodied and Lightly Armored Faunas of the Burgess Shale and of Some Other Classical Localities*, American Journal of Science. 301 (8).
44 Stephen C. Meyer, 2013. *Darwin's Doubt: The Explosive Origin of Animal Life*. San Francisco: Harper One.
45 Kuroda, P. K., 1956. On the Nuclear Physical Stability of the Uranium Minerals. Journal of Chemical Physics. Volume 25 (4): 781–782, 1295–1296.

Oklo, Gabon. Assays showed that the uranium ore body at the Oklo mine had deficiencies in the quantities of ^{235}U and discrepancies in neodymium (^{142}Nd, ^{143}Nd) and ruthenium (^{99}Ru) isotopes leading to an analysis and ultimately the confirmation that a reactor had indeed sustained periodic fission reactions for a period on the order of thousands of years. In my years as a geochemist at the Nuclear Regulatory Commission, we studied Oklo as the **natural analogue** of an accidental release of radionuclides. We were interested in rapid groundwater transport of isotopes like technetium (^{99}Tc) that decays to ruthenium (^{99}Ru) because of their extreme mobility in the Earth's crust and their ability to rapidly travel significant distances. Thus, swiftly transported, exceedingly long-lived radioisotopes, like ^{99}Tc and iodine (^{129}I) are important in the study of human health effects. They continue to cause a radiation dose long after hot fission products like cesium (^{137}Cs) have decayed. Recently, a number of scientists have proposed that the Earth may have similar natural nuclear reactors resident at the core-mantle boundary. The hypothesis was initially considered as a potential mechanism for a nuclear explosion that would blast the Moon from the ribs of the Earth.

> *...The fission hypothesis is an alternative explanation for the formation of the moon, and it predicts similar isotope ratios in the Moon and Earth... The hypothesis (credited to Charles Darwin's son George in 1879) is that the Earth and Moon began as a mass of molten rock spinning rapidly enough that gravity was just barely greater than the centrifugal forces. Even a slight kick could dislodge part of the mass into orbit, where it would become the Moon. The hypothesis has been around for 130 years, but was rejected because no one could explain a source of the energy required to kick a moon-sized blob of molten rock into orbit. The researchers suggest the hypothesis explains the identical isotopic composition of light and heavy elements, and further propose it could be tested, since the explosion would leave evidence such as xenon-136 and helium-3, which would*

have been produced in abundance in the georeactor.[46] ...the dynamics of this scenario requires on the order of 10^{29}–10^{30} J almost instantaneously generated additional energy if the angular momentum of the proto-Earth was similar to that of the Earth–Moon system today. The only known source for this additional energy is nuclear fission... it is feasible to form the Moon through the ejection of terrestrial silicate material triggered by a <u>nuclear explosion at Earth's core–mantle boundary (CMB), causing a shock wave propagating through the Earth</u>. Hydrodynamic modelling of this scenario shows that <u>a shock wave created by rapidly expanding plasma resulting from the explosion disrupts and expels overlying mantle and crust material.</u> Our hypothesis straightforwardly explains the identical isotopic composition of Earth and Moon for both lighter (oxygen, silicon, potassium) and heavier (chromium, titanium, neodymium and tungsten) elements. It is also consistent with the proposed Earth-like water abundances in the early Moon, with the angular momentum and energy of the present-day Earth–Moon system, and with the early formation of a 'hidden reservoir' at Earth's CMB that is not present in the Moon.[47]

Could the Earth accumulate natural nuclear reactor materials due to the differentiation processes of the Earth's Core and Mantle? Of course, the ejection of the moon could simply be the product of a ginormous collision with a small to medium size planet.

Day, Summer, 2300 BC, New Siberian Islands

The Artic meadows stretched as far as the eyes could see.[48] The meadows looked like a modern day alpine savanna with wildflowers, grasses, shrubs, and sedges covering the ground.

---

46 Edwards, Lin. 2010. The Moon may have formed in a nuclear explosion. phys.org.
47 de Meijer, R.J. and others. 2013. Forming the Moon from terrestrial silicate-rich material. Chemical Geology, Volume 345. Pages 40-49.
48 Reconstruction of possible event sequence by the author.

Flowers that looked like buttercups, poppies, and anemones[49] covered the ground with a beautiful carpet for the immense herds that foraged there. Giant herds of wooly mammoths were peacefully grazing in the warmth of the meadows in clear site of the giant forests that surrounded them, while small numbers of wooly rhinoceros were comfortably grazing in the distance. A large herd of horses, several musk ox, reindeer, and even bison roamed in harmony among the endless herds of mammoths. Sunshine stretched across the meadows adding to the serenity of the environment. But everything was about to change. Suddenly, flashes of light streaked across the sky as though the sun had exploded. The light was blinding, brighter than the streaking light over Chelyabinsk in 2013. The streaks of light were followed by loud sonic booms. The light, sound, and heaving ground like waves were followed by loud animal screams from the stampeding herds. Tiny fragments of raining fire struck animals within the herds,[50] adding to the frantic flight for life.

Towering clouds of ash jetted high into the sky and the daylight suddenly became eerily dark as the ash blotted out the sun. The smell of sulfur and other gases gave off the distinctive stench of smoke. Moments later, gale force winds raised havoc within the herds, separating the young from the adults. The heard felt the rumbling and ran for high ground as the earth beneath their hooves vibrated, groaned, and became unstable. But run as they may, it was no use. Predators mingled with their prey in the desperate flight to safety. And yet, the worst was soon to come. In the distance, a towering wall of water raced toward them and engulfed them. Enormous trees snapped like twigs and floated as enormous rafts floundering in gigantic waves as the tsunami sped across the landscape. All but a few met their fate in minutes, drowned by the tidal wave. Survivors of the onslaught soon succumbed to their watery grave.

---

49 Chung, Emily, 2014. Wooly mammoth diet mystery solved by DNA analysis. CBC News.
50 Jonathan T. Hagstruma, Richard B. Firestone , Allen West, Zsolt Stefanka and Zsolt Revay. 2010. Journal of Siberian Federal University. Engineering & Technologies pages 123-132.

As time progressed, piles of the dead were scattered across the Arctic. Carcasses of the wooly rhinoceros are found as far away as the British Isles and as far east as the Chukotka Peninsula, at the eastern tip of Siberia.[51] Many of the dead still have fleshly, albeit frozen, remains to this very day. As was the case following the volcanic episode at Tambora, the Earth's temperatures plummeted. Instead of a single volcano, hundreds of volcanoes were erupting and Planet Earth fell into the grip of an extended glacial period that preserved many of the animal and plant remains. Many are still buried in the enormous muck flows that were carried by the pulsating tsunamis that buried the dead.[52]

Today the thawing permafrost is yielding up the dead including buried forests.[53, 54] Plants and long forgotten deadly viruses are also being brought to life after millennia of burial.[55, 56] The plausibility of the above scenario is based on the geological evidence, the current state of the dead, and annals of similar events. The records of modern man document one catastrophic event after another. Events like the catastrophic eruption of Mount Vesuvius and the burial of the population at Pompeii; the prophetic Lisbon earthquake of November 1st, 1755, Feast of All Saints (Revelation 6:12); or the stone carvings on the pillars of the Temple at *Göbekli Tepe,* on the border of modern Turkey and Syria, not far from the post-Flood home of Noah; are just a few examples of numerous catastrophes that have plagued our Planet. But what about numerous catastrophes found in the record

---

51 Fleming, Nic, 2014. Four amazing mummified animals from the Ice Age. BBC. Earth.

52 Solovyov, Dmitry, 2007. Receding permafrost is a bone hunter's bounty. Reuters. Environment.

53 Macdonald, Fiona. 2018. There's a "Doorway to the Underworld" in Siberia so Bud it's Uncovered Ancient Forests. Sciencealert.com.

54 Pelley, Scott, 2019. Siberia's Pleistocene Park: Bringing Back Pieces of the Ice Age to Combat Climate Change. 60 Minutes.

55 Yashina, S. et al., 2012. Regeneration of whole fertile plants for 30,000-y-old fruit tissue buried in Siberian Permafrost. PNAS 109 (10) 4008-4013.

56 Alex, Bridget, 2019. Artic Meltdown: We're Already Feeling the Consequences of Thawing Permafrost. Discover Magazine.

inscribed in Nature's rocks? Do they record a record of uniformitarianism or catastrophism? You be the judge.

UNLOCKING EARTH'S APOCALYPTIC SECRETS

Of old hast thou laid the foundation of the earth: and the heavens [are] the work of thy hands[57] ... the works ...finished from the foundation of the world.[58] If thou seekest her as silver, and searchest for her as [for] hid treasures; Then shalt thou understand...For the LORD giveth wisdom: out of his mouth [cometh] knowledge and understanding.[59] The invisible things of him from the creation of the world are clearly seen[60]... the treasures of darkness, and hidden riches of secret places...[61]

Independent, unconnected, catastrophic events are insufficient to topple Uniformitarianism. Is Uniformitarianism a firm foundation for Evolution or is it sinking sand? Scientific investigation provides the methods for detecting the synergistic building blocks of an uninterrupted Apocalyptic Pattern ever since the Cambrian Explosion of life. If proven, such a pattern would undermine Evolution. Like unlocking a complex enigma, code, or puzzle we begin by unearthing the hidden secrets of the Earth's genetic code of catastrophism.

---

57 Psalm 102:25
58 Hebrews 4:3.
59 Proverbs 2:3-6.
60 Romans 1:20
61 Isaiah 45:3.

2 *The Rocks Cry Out*

And I will give thee the treasures of darkness, and hidden riches of secret places[62]… the mystery which hath been hid from ages and from generations, but now is made manifest…[63] For the stone shall cry out [64] …and there shall be a time of trouble, such as never was since there was a nation [even] to that same time…[65] Behold, waters rise up out of the north, and shall be an overflowing flood, and shall overflow the land, and all that is therein;[66] the heavens shall pass away with a great noise, and the elements shall melt with fervent heat, the earth also and the works that are therein shall be burned up.[67]

At daybreak, in July of 1973, I loaded my old green 1967 F-100 Ford pick-up, left camp, and headed from Cotopaxi, Colorado to my research area between Lookout Mountain and McClure Mountain. I had been systematically mapping a 200 square mile area south of the Arkansas River. On this morning, I drove east along the mighty Arkansas River Gorge with its exposures of pink granites and crossed at Texas Creek and headed south along highway 69. I was unaware that it would be a day filled with a crucial discovery and a close brush with the wild.

A picture was emerging that tied a series of global Cambrian events together; violent magmatic intrusions accompanied the sudden appearance of life on Planet Earth referred to as the **Cambrian Explosion.** But there was more, much more. As the

---

62 Isaiah 45:3.
63 Colossians 1:26.
64 Habakkuk 2:11.
65 Daniel 12:1.
66 Jeremiah 47:2.
67 2 Peter 3:10.

years flew by, a frightening picture of the catastrophes of the past and their implications for the future began to emerge. *What would be the events leading to the End of Time: even of mankind?*

As I drove South, I admired the beauty of the majestic **Sangre de Cristo (Blood of Christ)** Range to the west. The Sangre de Cristo Range is a block faulted range, much like the Teton Range, and rises as much as 7000 feet above a wide valley. They were lifted by extremely powerful collisions between two ancient continents. But there was a strange void of any younger rocks except recent gravels; leading to the conclusion that a staggering amount of sediments on the order of <u>kilometers,</u>[68] <u>representing as much as a billion year time gap,</u>[69] <u>had been eroded away</u>, exposing the rare Cambrian rocks. It was the **Great Unconformity** that could be traced to the Grand Canyon and around the globe. The Cambrian was a time when the supercontinent of Rodinia was breaking into fragments and a time when the Sauk seas raged over the continents and this very valley. I couldn't help but think that Rodinia's larger fragments, Laurentia and Gondwana collided near this place. I was sure **the driving forces of the collisions were associated with the emplacement of the rare Cambrian rocks our team was studying.**

Turning East along a familiar dirt road that led to the ancient magmatic intrusions, I pulled off a side road and parked at the foot of a high ridge and put on my pack, and grabbed a hammer, a field hat, and a bagged lunch. The ascent was fairly severe and I knew that I would bring more than 50 pounds of rock samples from the ridge if I found the rare Cambrian rocks. The path that I followed was densely covered with piñon pines and the overgrowth made it harder to find samples. I could hear the crunch of gravel as I ascended the steep trail. As I neared the crest, my eye spotted fragments (called float) of a strange rock, and I followed them like a trail of breadcrumbs. My excitement grew with each step but

---

68 Heinrich, E. Wm. And D.H. Dahlem. 1969. Dikes of the McClure mountain — Iron mountain alkalic complex, Fremont county, Colorado, U.S.A. Bulletin Volcanologique volume 33, pages960–976.
69 Lindsey, David. A. 2010. The Geologic Story of Colorado's Sangre de Cristo Range. USGS Circular 1349.

booming claps of thunder drew my attention from my survey and I knew that I would need to get back to the main road before the torrential downpour transformed the road to mud. But I was so, so very close that I decided to take a chance and complete my climb to the ridge and take samples. I wouldn't be disappointed. There, near the ridge, I found an exceptional rock formation.[70] As I looked upon the rock I envisioned swirling, rushing magma and gases exploding through the older granites of the valley. The rocks provided evidence of liquid immiscibility like a solution of vinegar and oil. You can see the flow of two liquids, frozen in time, in this photo of a 12 inch wide rock slab taken from an igneous dike.

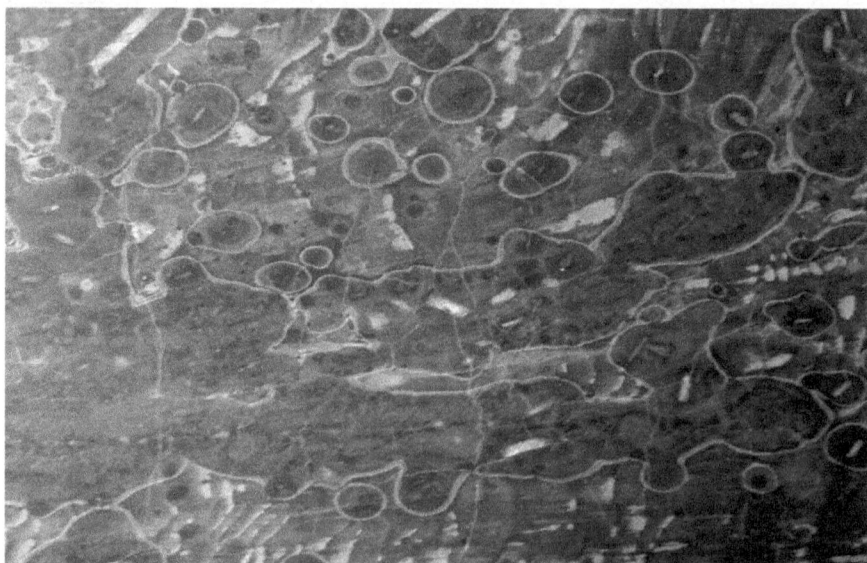

Liquid immiscibility is a particularly attractive petrogenetic model, given the apparent bimodal distribution of rock types at numerous alkaline silicate–carbonatite complexes... and is currently the favoured mechanism for a number of localities including Oldoinyo Lengai (Mountain of God)...[71]

---

70 Alexander, D.H. 1974. Petrography and origin of an orbicular lamprophyre dike, Fremont County, Colorado. U. Michigan Master's Dissertation. 106 pp.
71 Brooker, R.A. and B.A. Kjarsgaard. 2011. Silicate–Carbonate Liquid Immiscibility and Phase Relations in the System SiO_2–Na_2O–Al_2O_3–CaO–

The magmas that formed these rocks were violently injected in fractures and faults, propelled by supercritical fluids of water,[72] carbon dioxide, methane, sulfur and fluorine gases that opened the granitic fractures. Gases and fluids created a halo around the complex, bearing exotic elements of thorium, uranium, and Rare Earth elements (like gadolinium and neodymium). These types of intrusions are related to the kimberlite pipes of Africa, made famous for their high concentrations of diamonds. And diamonds provide a window into the Earth's mantle. I wondered if there was a connection among the kimberlites, lamproites, and carbonatites of Canada, Colorado, southern Argentina, and the Cambrian kimberlites of Africa,[73] all of Cambrian age. Afterall, these landmasses were once welded together within the supercontinent of Rodinia. But perhaps there was a more ominous explanation for the intrusion of these thorium and uranium rich veins...[74]

In my excitement, I had forgotten about the oncoming storm, until sudden flashes of light and thunder grew near, and the temperature rapidly dropped. I quickly numbered each sample, pinpointed their location on my map, and filled the pack and headed down the trail. On the descent I was still scanning the ground for additional evidence when a chill ran up my spine. Tracks of a large mountain lion had been stalking me and he left his enormous pawprints on my boot tracks. I grimaced, clutched my hammer, and slammed it on the rocks to frighten the cougar away. Ironically, I was only a mile or two from a mountain lion outfitter's camp. Over the early field seasons I caught glimpses of cougars, rattlers, and friendlier game; even unwittingly driving two cougar cubs up a tree. Needless to say, it was time for a quick retreat because their den and their mother were nearby. The work of my colleagues and

---

CO_2 at $0 \cdot 1$–$2 \cdot 5$ GPa with Applications to Carbonatite Genesis. Journal of Petrology, Volume 52, Issue 7-8, Pages 1281–1305.

72 Galli, G. and Ding Pan. 2013. A closer look at supercritical water. PNAS. Volume 110. Pp. 6250-6251.

73 Torsvik, T.H., and others. 2010. Diamonds sampled by plumes from the core–mantle boundary. Nature. Volume 466. Pp. 352-355 plus supplements.

74 Edwards, Lin. 2010. The Moon may have formed in a nuclear explosion. phys.org.

predecessors[75] and the specimens that I collected over four field seasons provide a key piece of evidence for unlocking the early Cambrian events surrounding the *Cambrian Explosion of Life. The rocks have a story to tell about the past and your future.*

Cambrian Opening of the Early Atlantic

Over the next several years, I returned to the field area to complete my extensive sampling project. During one special summer, I collected spatially oriented samples for paleomagnetic work. Our objective was aimed at evaluating *the opening of the Cambrian proto-Atlantic ocean that is called Iapetus*:

> *In recent years, the hypothesis by Wilson (1966) that a proto-Atlantic ocean existed between Europe and North America during early Paleozoic time has steadily gained acceptance...The Iron Mountain-McClure Mountain complex has been extensively studied by Parker and Hildebrand (1963), Shawe and Parker (1967) and Heinrich and Dahlem (1969). It is the subject of ongoing investigations by Heinrich and Alexander (University of Michigan) and by the U.S. Geological Survey...the samples were collected from 35 sites, and the samples were demagnetized using thermal and alternating-field techniques...the two methods yield very similar directions...consistent with published Cambrian poles...[76]*

Fossil evidence on both sides of the Atlantic have led a number of workers to provide additional evidence for the correlation of distant coastlines. The famous paper by J. Tuzo Wilson added a degree of refinement to our understanding, making the case for the opening and closing of ocean basins and rifts:

---

75 Heinrich, E. Wm. And D.H. Dahlem. 1969. IBID.
76 R.B. French, D.H. Alexander, and R. Van der Voo. 1977. Paleomagnetism of upper Precambrian to lower Paleozoic intrusive rocks from Colorado. GSA Bulletin Volume 88 (12). Pages 1785-1792.

It was Tuzo Wilson (1966) who recognised that the different faunal distributions on both sides of the present-day North Atlantic Ocean required the existence of an earlier proto-Atlantic Ocean. The observation that the present-day Atlantic Ocean mainly opened along a former suture was a crucial step in the formulation of <u>the Wilson Cycle theory</u>. The theory implies that collision zones are structures that are able to localize extensional deformation for long times after the collision has waned... For margins that are associated with large igneous provinces (LIPs), we find a positive correlation between break-up age and LIP age. We interpret this to indicate that plumes can aid the factual continental break-up. However, plumes may have been guided towards the rift for margins that experienced a long rift history (e.g., Norway-Greenland), to then trigger the break-up. This could offer a partial reconciliation in the debate of a passive or active role for mantle plumes in continental break-up. (Wilson, J.T., 1966. Did the Atlantic close and then re-open? Nature 211, 676-681)[77]

Continental Collisions on Proto-Atlantic Shores

At the time life was emerging, the supercontinent of Rodinia was breaking up into fragments. Catastrophic rifting and collisions were occurring at the time of the **Cambrian Explosion**. These mountain building events were occurring among the former shores of Rodinia's fragments including Laurentia (the East Coast of North America), Greenland, and Europa:

The Central Appalachians of eastern Pennsylvania... display a complete geological record of a long-lived evolution from the late Neoproterozoic rifting of the Rodinia supercontinent to the Middle– Late Ordovician collision between Laurentian craton and an assemblage of microcontinents and magmatic arc units that culminated in the Taconic Orogeny ... After the dismantling of Rodinia (~620 to 580 Ma) a line of

---

77 Buiter, S. and T. Torsvik. 2013. Geophysical Research Abstracts Vol. 15, EGU2013-2596.

microcontinents (i.e., Brandywine and Baltimore Terranes) was separated from Laurentian craton to the NW by a seaway of the western Iapetus Ocean ... In the late Early Ordovician, this Octoraro Sea began closing as a result of the subduction[78] of its seafloor eastward beneath the archipelago of these rifted microcontinents and a magmatic arc system. This convergence and the associated subduction zone tectonics produced a northward advancing accretionary wedge ... and resulted in the collision of the microcontinents with the Laurentian margin in the down-going plate during the Late Ordovician...[79]

Imagine, an entire ocean floor disappearing into the mantle beneath another continent! What forces were involved? In 1969, I was on a field trip in southern New York State where we viewed massive layered rock beds of the northern Appalachian basin. Apparently, the **continental collisions were so powerful that they sent muddy slurries and even mountainsides (klippes) westwards traveling at high speeds.** At one particular outcrop, we were given a description of the large horizontally layered rock unit. The unit was a Bouma sequence[80] and I was amazed when we learned that the same rock unit extended beyond the western border of the state, some hundreds of miles to the west. We were told that the bed was a turbidite[81] that had been deposited rapidly underwater like the Newfoundland turbidite that raced across the Atlantic ocean.

---

78 Oskin, Becky. 2015. What Is a Subduction Zone? Livescience.com. A subduction zone is the biggest crash scene on Earth. These boundaries mark the collision between two of the planet's tectonic plates. The plates are pieces of crust that slowly move across the planet's surface... Where two tectonic plates meet at a subduction zone, one bends and slides underneath the other, curving down into the mantle.

79 Codegone, G. and others. 2012. Formation of Taconic mélanges and broken formations in the Hamburg Klippe, Central Appalachian Orogenic Belt, Eastern Pennsylvania. Tectonophysics. Volumes 568-569. Pages 215-229.

80 Bouma, Arnold H. (1962). *Sedimentology of some Flysch deposits: A graphic approach to facies interpretation.* Elsevier. p. 168 p

81 Lowe, D.R. (1982). "Sediment gravity flows: II. Depositional models with special reference to the deposits of high-density turbidity currents". *Journal of Sedimentology.* Society of Economic Paleontologists and Mineralogists: v. 52, p. 279–297

The **Cambrian Explosion of life** on the fragmented shores of Rodinia found in fossil localities like the Burgess Shales of British Columbia, Morocco, Chenjiang County in China, the Aldan River site of Siberia, the Emu Bay shale of Australia, and the Sirius Passet site in Greenland among others, was accompanied by the catastrophic emplacement of the unusual magmas I studied in Colorado, and continental collisions along the East Coast of North America. Without question, the Earth was in turmoil and upheaval at the critical juncture when life suddenly emerged. Cambrian fossil localities are found in the House Range of Utah, and in the Marble Mountains and the White-Inyo Mountains of California. And when the Iapetus Ocean opened as Rodinia broke apart, small rift basins formed in New York. Yet there is much more to this volume of the **Earth's Chronicles**.

Cambrian Addy Sandstone at the Western Continental Margin of North America

Along the western margin of the United States another Cambrian coastline was in the making. North of Spokane, the distinctive Addy Sandstone marks the Cambrian coastal zone and the sandstone, products of great floods and likely glacial scouring:

> **The great supercontinent of Rodinia dominated the Earth ... The one great planetary ocean undoubtedly spawned violent storms over its expansive waters. With nothing to protect the landscape from the forces of water, great floods were probably common on the continent...But, not even supercontinents last forever. In the end, Rodinia fell victim to the Earth's internal heat. <u>A slow buildup of heat beneath Rodinia caused the old continent's crust to dome, stretch and weaken. Eventually, the entire continent ruptured.</u> Violent spreading centers developed underneath the continent and began to slowly tear Rodinia apart... The rupture that split Rodinia abruptly truncated the basin in which the**

Belt sediments accumulated. ...The broken edge of Rodinia through eastern Washington was a quiet, passive continental margin, far removed from the violent tectonics we associate with the Pacific Northwest today...As the rift continued to grow, it eventually went on to form a vast ocean basin called the "Panthalassic Ocean," a Greek word for "all the seas." The Panthalassic Ocean separated the Americas, Siberia and Scandinavia from Antarctica, Australia, and the rest of the eastern hemisphere. The new coastline of ancestral North America now ran through what is today eastern Washington, not too far east of modern-day Pullman. As the Pacific Northwest comfortably passed through the Paleozoic Era, it witnessed a veritable explosion of life. Off its shores, most of the major groups of animals first appear in the fossil record in the Cambrian Period...As Rodinia split, the Panthalassic Ocean became ever wider. Undoubtedly, a mid-ocean spreading center lie along its axis creating new oceanic crust on the floor of the young ocean. [82]

Catastrophic Earth History

Over the years I have witnessed extensive evidence of a shattered Planet. In Sudbury Ontario, we witnessed, first-hand, the after effects of one of the world's largest asteroid/comet impacts at Sudbury Ontario. It reportedly landed in a sea environment with such force that it created magma and volcanoes and scattered debris over more than 1.5 million square kilometers; even to Minnesota and beyond. It's impact had an estimated force equal to several billion Hiroshima-sized atomic bombs; **_enough force to bring magma up from the mantle_**.

---

82 Burke Museum. Retrieved online 2019. Dance of the Giants. burkemuseum.org.

3 *Signs of an Apocalyptic Past*

And on that day, ... the whole earth [shook], and the sun darkened, and the foundations of the world raged, and the whole earth was moved violently, and the lightning flashed, and the thunder roared, and all the fountains in the earth were broken up, such as was not known to the inhabitants before...[83]

Imagine a time when life suddenly appeared almost instantaneously upon the Earth without any apparent evolutionary predecessors. Could it be that life arose by Intelligent Design? It was a time when nearly as many animal phyla appeared as exist today! Imagine a time, shortly after this explosion of life, when these animals were buried so rapidly that even their soft body parts were preserved. Imagine a time when the Earth's surface was in such upheaval that entire land masses were moved vast distances in rapid motion. It was a period when continents were breaking-up and reforming. Volcanic eruptions were accompanied by giant tsunamis. Low lying plains and rolling hills were rapidly changing by catastrophic plate tectonics, often triggered by destructive meteoritic impacts. Imagine a period of time when erosion was so severe that sediments, kilometers thick, accumulated as if the land were a piece of wood that had been sanded away. Mountain ranges rose along colliding continental coasts as catastrophic floods swept across the globe at a scale unheard of today. Turbidite bearing sediments laid down extensive sedimentary deposits, even across continents. Geologists refer to this period of worldwide flooding as the Cambrian Sauk transgression.

---

83 Jasher 6:11.

Shattered Planet: Signs of an Apocalyptic Past

Geologists have mapped the planet's crust and observe that it is shattered into some ~9 major plates and dozens of minor and micro plates: for comparison, roughly the thickness of an eggshell as compared to a whole egg. As the Earth spins, enormous hot cells in the mantle move surface plates and intermittently cause earthquakes spelling disaster.

What energizes these hot cells?

> *... plate tectonics and the Earth's magnetic field were the result of massive collisions during the "geologic dark age" Andrew Masterson reports. Meteorite impacts might have kick-started the Earth's tectonic plates and boosted the planet's magnetic field... Modelling by O'Neill[84] and his colleagues has thrown up a possible answer. "Our results indicate that giant meteorite impacts in the past could have triggered events where the solid outer section of the Earth sinks into the deeper mantle at ocean trenches – a process known as subduction... This would have effectively recycled large portions of the Earth's surface, drastically changing the geography of the planet. We've seen evidence of some geological activity that suggests something like subduction acted on the early Earth... Meteorite impacts...would have caused the planet's cold outer crust to move comparatively rapidly downwards towards the planetary core. This process would have changed the intensity of convection within the core, thus affecting the "geodynamo" – the conductive layer of liquid iron that surrounds the solid inner core and that generates a magnetic field.[85]*

---

84 C. O'Neill et al. Impact-driven subduction on the Hadean Earth, *Nature Geoscience* (2017).
85 Andrew Masterson. COSMOS, The Science of Everything. September 26, 2017. Did meteorites create the Earth's tectonic plates?

CAMBRIAN: *Buried Lifeforms, Mega-Impacts, Continental Obliteration, and Epic-Flooding*

Deep within the Earth's sediment pile, in what scientists refer to as the Cambrian, are profound hidden mysteries of the origin of life. For it is here that a sudden explosion of many of today's existing phyla of animal life suddenly appear. It is here that geochemists have uncovered a significant rise in the planet's oxygen levels. And there is much, much-more to this story. According to a report in the July 1997 issue of SCIENCE by Kirschvink and others:

> *The Vendian-Cambrian transition [~600 to 500 million years ago (Ma)] is one of the most intriguing periods in Earth history. Geological evidence points to the breakup of one supercontinent, Rodinia, and the almost simultaneous assembly of another, Gondwanaland (1). The sudden appearance of virtually all the animal phyla (2) and their exponential diversification are coeval with abrupt shifts in oceanic geochemistry (3, 4). Recent calibration of this time interval with U-Pb isotopic ages (5, 6) indicates that these events occurred within a span of 30 million years (My), and <u>the major diversification happened in only 10 to 15 My (Fig. 1). The new ages, along with paleomagnetic data, indicate that continents moved at rapid rates that are difficult to reconcile with our present understanding of mantle dynamics</u> (7).* [86]

Perhaps most startling is the claim by Kirschvink and others that the plates were shifting on grand scales:

> *at least two tectonic plates, involving more than two-thirds of Earth's continental lithosphere, were involved in <u>a rapid rotation of ~90° relative to the spin axis</u>. We speculate that the entire lithosphere may have been involved in this*

---

[86] Kirschvink, Joseph, L. Robert L. Ripperdan, and David A. Evans. July 1997. Evidence for a Large-Scale Reorganization of Early Continental Masses by Inertial Interchange True Polar Wander. SCIENCE. Vol. 277. Pp. 541-545.

rotation. The postulated event began sometime in the first half of the Early Cambrian and ended by earliest Middle Cambrian time...

Imagine whole continents rapidly rotating 90° relative to the Earth's spin axis! Recent studies continue to draw the connection between rapid tectonic plate motion,[87] widespread volcanic activity, and the Cambrian explosion. Other authors have attributed this massive tectonism during the Cambrian to the sudden rise in oxygen levels.[88]

Recent work indicates a large increase in the tectonic CO_2 degassing rate between the Neoproterozoic and Paleozoic Eras. We use a biogeochemical model to show that this increase in the total carbon and sulphur throughput of the Earth system increased the rate of organic carbon and pyrite sulphur burial and hence atmospheric pO_2. Modelled atmospheric pO_2 increases by ~50% during the Ediacaran Period (635–541 Ma)...

Giant Asteroids Splintered Earth's Crust

Likewise, recent studies by Stanford Researchers, Norman Sleep and Donald Lowe, conclude that an ancient asteroid impact kick-started plate tectonics.[89] They claim to have discovered evidence of a giant impact in the Barberton greenstone belt of South Africa. They postulate that the asteroid was at least 23 miles wide and punched a crater nearly 500 kilometers (300 miles) wide. However, Chatterjee's research suggests that a similar sized asteroid formed the Shiva crater at the time of the dinosaurs. Some believe that the Shiva crater is merely the consequence of

---

87 Mitchell, Ross N and others. 2015. Was the Cambrian Explosion Both and Effect and Artifact of True Polar Wander? American Journal of Science, Vol. 315. Pages 945-957.
88 Williams, Joshua J., Benjamin J, W. Mills, and Timothy M. Lenton. 2019. A Tectonically Driven Ediacaran Oxygenation Event. Nature Communications, Vol. 10, Article 2690.
89 Bompey, Nancy. April 2014. Scientists Reconstruct Ancient Impact that Dwarfs Dinosaur-Extinction Blast. https://pangea.stanford.edu/news/scientists-reconstruct-ancient-impact-dwarfs-dinosaur-extinction-blast

subsidence due to oil drilling in the region. Another study by Sara Mazrouei and others[90] conclude that the frequency of Earth's impacts can be inferred by studying lunar craters. The research of this third team of geoscientists concludes that ***the impact rate greatly increased post-Cambrian on Earth*** *as it did on the Moon*.

Accelerated Plate Motion

The Earth has been punctuated by large asteroid impacts which, based on the above studies, were key to kick-starting periods of rapid motion of the Earth's giant plates. These periods of rapid plate motion likely triggered a ***massive surge in volcanic activity and even mass extinctions***. According to Byrnes and Karlstrom:[91]

> *Eruptive phenomena at all scales, from hydrothermal geysers to flood basalts, can potentially be initiated or modulated by external mechanical perturbations. We present evidence for the triggering of magmatism on a global scale by the Chicxulub meteorite impact at the Cretaceous-Paleogene (K-Pg) boundary, recorded by transiently increased crustal production at mid-ocean ridges. Concentrated positive free-air gravity and coincident seafloor topographic anomalies, associated with seafloor created at fast-spreading rates, suggest volumes of excess magmatism in the range of ~10^5 to 10^6 km^3. Widespread mobilization of existing mantle melt by post-impact seismic radiation can explain the volume and distribution of the anomalous crust. This massive but short-lived pulse of marine magmatism should be considered alongside the Chicxulub impact and Deccan Traps as a contributor to geochemical anomalies and environmental changes at K-Pg time.*

90 Mazrouei, Sara et al. January 2019. Earth and Moon Impact Flux increased at the end of the Paleozoic. Science, Vol. 363, Issue 6424, pp. 253-257.
91 Byrnes, Joseph S. and Leif Karlstrom. 2018. Anomalous K-Pg-aged Seafloor Attributed to Impact-Induced Mid-Ocean Ridge Magnetism. Science Advances, AAAS.

Similarly, Andrew Glikson[92] concludes that periodic impact clusters and accelerated volcanism are associated with extinctions:

> *...both extraterrestrial impacts and volcanism served as extinction triggers separately or in combination. A protracted impact cluster overlaps extinctions at the end-Devonian (~374–359 Ma) and impact-extinction age overlaps occur in the end-Jurassic (~145– 142 Ma), Aptian (~125–112 Ma); Cenomanian–Turonian (~95–94 Ma); K–T boundary (~65.5 Ma) and mid-Miocene (~16 Ma).*

WORLDWIDE FLOODS: Invasion of the Seas

Perhaps the greatest clue to the mystery of the Cambrian was reported in numerous articles that address the missing impact craters of the Earth. Robin Andrews, a writer for National Geographic summarizes the scientific findings as follows:[93]

> *For one thing, "around 600 to 700 million years ago, Earth loses its craters," notes study coauthor Bill Bottke, a planetary scientist and asteroid expert at the Southwest Research Institute in Boulder, Colorado. Some ancient craters still exist on stable continental cores named cratons, but they are few and far between...The easy explanation for this mystery was also <u>a ginormous erosional event,</u> but until now, evidence for one was hard to come by. Unlike many other worlds, "Earth does a really good job at erasing the tracks of its past," Bottke says. Fortunately, Keller's geochemistry made it clear that Snowball Earth provides a natural explanation. Then there's the huge uptick in sedimentation rates at the start of the Cambrian. All the*

---

92 Glikson, Andrew. 2005. Asteroid/Comet Impact Clusters, Flood Basalts and Mass Extinctions: Significance of Isotopic Age Overlaps. Earth and Planetary Science Letters 236 (2005) 933-937.
93 Robin George Andrews. 2018. Earth is Missing a Huge Part of Its Crust. Now we may know why. National Geographic.
www.nationalgeographic.com/science/2018/12/part-earths-crust-went-missing-glaciers-may-be-why-geology/

new sediment required plenty of space to fall into, something that would have only been possible if massive levels of erosion took place beforehand, says coauthor Thomas Gernon, an associate professor of earth science at the University of Southampton…As the researchers point out, one problem with their data is that there is still a multimillion-year time gap between the predicted end of Snowball Earth and the start of the Cambrian. It's not clear why the formation of new rock layers took so long to start after all that erosion stopped.

Perhaps there is a far better explanation for massive erosion at the beginning of the Cambrian. Glacial scouring combined with flooding when sea levels were high and the land elevations were low were the likely mechanistic culprits of deep erosion. Geologists call this period of dramatic global flooding, the Sauk transgression. According to Karl Karlstrom and coauthors:[94]

When linked to calibrated trilobite zone ages of greater than 500 million years old, these age constraints show that the marine transgression across a greater than 300-km-wide cratonic region took place during an interval 505 to 500 million years ago—more recently and more rapidly than previously thought. We redefine this onlap as the main Sauk transgression in the region. Mechanisms for this rapid flooding of the continent include thermal subsidence following the final breakup of Rodinia, combined with abrupt global eustatic changes driven by climate and/or mantle buoyancy modifications.

Global floods occurred in the Cambrian and reoccurred at regular intervals all the way to the present day. The last period of global flooding ended with the retreat of the Tejas at about 23 million years before present. Are we soon to be flooded for a 7th and final time?

---

94 Karlstrom, K et al. May 2018. Cambrian Sauk transgression in the Grand Canyon region redefined by detrital zircons. Nature Geoscience. Vol. 11, pages 438-443.

4 *Rodinia Break-up, Mega-Floods & a Lost Ocean*

He set a compass upon the face of the depth: When he established the clouds above: when he strengthened the fountains of the deep: When he gave to the sea his decree, that the waters should not pass his commandment: when he appointed the foundations of the earth:[95] *God is our refuge and strength, A very present help in trouble. Therefore, we will not fear, though the earth should change And though the mountains slip into the heart of the sea; Though its waters roar and foam, Though the mountains quake at its swelling pride.*[96]

Forces, almost incomprehensible to the human experience, so extreme, so powerful, so far beyond the imagination of mankind, have shaped and reshaped the planet we call home; Planet Earth. Recent revelations about our near twin planet Mars suggest that it once had an atmosphere and an ocean that have since been stripped away by Solar winds and other forces. For perspective, imagine for a moment the forces generated by the capture of the moon. Imagine how these forces would transform the world as we know it. As a minimum, the tidal forces would have caused heaving of the continents and the oceans. *By definition, the transformation would be of such an incredible magnitude, so violent, and so extensive that no uniformitarian theory could bring it about. Many say that global flooding is not possible but the Rock Record provides evidence of periods of frequent global flooding on Earth.*

---

95 Proverbs 8:27-29.
96 Psalm 46:1-10.

"Mars appears to have had a thick atmosphere warm enough to support liquid water which is a key ingredient and medium for life as we currently know it," said John Grunsfeld, astronaut and associate administrator for the NASA Science Mission Directorate in Washington. "Understanding what happened to the Mars atmosphere will inform our knowledge of the dynamics and evolution of any planetary atmosphere. Learning what can cause changes to a planet's environment from one that could host microbes at the surface to one that doesn't is important to know, and is a key question that is being addressed in NASA's journey to Mars."[97]

NASA is under direction to take the first steps toward the colonization of Mars by the 2030's. Is there a hidden agenda?

Devastating Megatsunamis

The tallest recorded tsunami wave was at Lituya Bay, Alaska in 1958, largely because it was confined to a narrow channel and reportedly reached as high as 524 meters (1720 feet). But observers say the wave was more like a hundred feet but the splash uprooted forests as high as 1720 feet. Mr. Howard Ulrich and his son Eddie had been boating in Lituya Bay on the evening of July 9, 1958.[98] At around 9:00 PM a 7.7 magnitude earthquake along the Fairweather Fault caused a rock slide of about 40 million cubic yards drop from a cliff high above the northeastern shore of Lituya Bay. The rock slide plunged from an elevation of approximately 3000 feet down into the waters of Gilbert Inlet. The impact force of the rockfall generated a local tsunami that crashed against the southwest shoreline of the inlet. Millions of trees were uprooted and swept away by the wave. Devastating tsunamis have killed

---

97 NASA, November 5, 2015. NASA Mission Reveals Speed of Solar Wind Stripping Martian Atmosphere. www.nasa.gov/press-release/nasa-mission-reveals-speed-of-solar-wind-stripping-martian-atmosphere
98 Geology.com Posted July 18, 2019. World's Tallest Tsunami: A tsunami with a record run-up height of 1720 feet occurred in Lituya Bay, Alaska. Geoscience News and Information.

hundreds of thousands of people over the past several centuries, destroying forests, causing rock slides, displacing villages, destroying cities, and even disabling a nuclear power plant. All told, these tsunamis resulted in hundreds of billions of dollars in damages. Yet the majority of the waves ranged in height from about 6 meters (20 feet) to 50 meters (164 feet). The majority of historical tsunamis have been reported from the margins of the Pacific Ocean and are most prevalent in Japan and Indonesia. According to Australian Geographic the following are several of the most destructive tsunamis in recent history:[99]

Sumatra, Indonesia – 26 December 2004	*50 meters*
Fukushima, Japan – 11 March 2011	*10 meters*
Lisbon, Portugal – 1 November 1755	*30 meters*
Krakatau, Indonesia – 27 August 1883	*37 meters*
Nankaido, Japan – 28 October 1707	*25 meters*
Sanriku, Japan – 15 June 1896	*38 meters*
Northern Chile – 13 August 1868	*21 meters*

I encourage the reader to watch YouTube videos of the Fukushima tsunami to get a feel for the catastrophic devastation of even a 10 meter wave.[100]

Ginormous Mile High Splash

As devastating as the historically recorded tsunamis are, they are like small ripples in comparison to tsunamis of the Earth's past. Recent work by a team of scientists at the University of Michigan[101] suggests that the 9 mile wide Chicxulub dinosaur killing asteroid that struck the Yucatan sent a nearly **mile high**

99 Phillips, Campbell. March 16, 2011. The 10 Most Destructive Tsunamis in History. Australian Geographic.
www.australiangeographic.com.au/topics/science-environment/2011/03/the-10-most-destructive-tsunamis-in-history/
100 www.youtube.com/watch?v=w3AdFjklR50
101 Geggel, Laura. January 2019. Dinosaur-Killing Asteroid Triggered Mile-High Tsunami That Spread Through Earth's Oceans. LiveScience.
www.livescience.com/64426-dinosaur-killing-asteroid-caused-giant-tsunami.html

<u>**initial splash**</u> wave that sent tsunamis through the Gulf of Mexico and throughout the existing global oceans. Initial modeling by team member Brandon Johnson of Brown University indicated that the asteroid created a nearly mile deep crater. Waves from the impact exited the Gulf of Mexico through the Central America Seaway. The Gulf Coast saw waves ranging from 13 to 330 feet while in the North Atlantic waves reached 46 feet and those of the North Pacific reached 13 feet. The single asteroid sent tsunamis racing over continents around the world. But what is more startling, is that the wave height of the tsunami was limited by the comparatively shallow waters of the Yucatan. According to Bryant,[102] *<u>if the same asteroid were to impact the open ocean it would have created a wave of about 2.9 miles high.</u>* If such an asteroid were to strike the Gulf of Mexico today, it would have wiped out a number of states including Florida.

According to Joanne Bourgeois,[103] substantial evidence demonstrates that tsunamis can dramatically change the landscape: *tsunamis eroded and deposited not only sand, but also large boulders and coral debris.* Impact of the largest known asteroids in the Pacific Ocean for instance, would cause such a displacement that it would have likely caused obduction of the crust, such as is seen in the upward thrusting of the Himalayans and subduction as is seen in the case of the Farallon Plate slipping more than a thousand miles inland beneath the North American Plate causing the rise of mountains from California to Colorado. Megatsunamis and their associated turbidity currents would tear up the earth beneath the flow, much like the waves of water that dramatically carved the Columbia Basin as a consequence of the Missoula Flood. But whereas the Missoula Flood was confined to the Basin, these Megatsunamis would have a terrifying global impact.[104]

---

102 Bryant, Edward. June 2014. *Tsunami: The Underrated Hazard.* Springer. p. 177-178 . *ISBN 978-3-319-06133-7.*
103 Bourgeois, Joanne. 2009. Chapter 3. Geologic Effects and Records of Tsunamis. In, The Sea, Volume 15. Tsunamis. Harvard University Press, 2009. Pages 53-91.
104 Kornei, Katherine. December 2018. Huge Global Tsunami Followed Dinosaur Killing Asteroid Impact. EOS. Earth & Space Science News.

The Breakup of Rodinia and the Vanishing of the Iapetus Ocean

Scientists believe that a supercontinent referred to as Rodinia began to be aggressively broken apart during the Cambrian. The continent was thought to be near the South Pole based upon paleomagnetic reconstructions[105] and ultimately split into two continents known as Gondwana and Laurentia and other lesser fragments. Imagine what incredible forces it would take to split apart a continent like North America! Some possible causes and details of the breakup of Rodinia are discussed in a study by Bogdanova and others on behalf of UNESCO.[106] The scientists suggest that a superplume[107] of molten magma caused the lighter continental crust to arch and break along rift zones which were the centers of extensive lava flows and volcanic eruptions. Is it possible that this superplume resulted from an asteroid impact?[108]

In the meantime, an associated set of events is thought to have led to the opening of the lost Iapetus Ocean which formed between Laurentia and two other fragments of Rodinia referred to as Baltica and Avalonia. As noted earlier, Iapetus represents the predecessor to the Atlantic Ocean. Clarkson and Upton, in their book, **Death of an Ocean: A Geologic Borders Ballad**[109] detail how the

https://eos.org/articles/huge-global-tsunami-followed-dinosaur-killing-asteroid-impact

105 A. B. Weil, R. Van der Voo, C. MacNiocaill, and J. G. Meert, "The Proterozoic Supercontinent Rodinia: Paleomagnetically Derived Reconstructions for 1100 to 800 Ma," Earth Planet. Sci. Lett. **154**, 13–24 (1998).

106 Bogdanova, S. V.; Pisarevsky, S. A.; Li, Z. X. (2009). "Assembly and Breakup of Rodinia (Some Results of IGCP Project 440)". *Stratigraphy and Geological Correlation.* **17** (3): 259–274.

107 Z.X Lia, X.H Li b P.DKinny c J Wang d.1999. The Breakup of Rodinia: did it start with a mantle plume beneath South China? Earth and Planetary Science Letters. Vol. 173, Issue 3. Pp. 171-181.

108 Citron, Robert I, Michael Manga, and Eh Tan. 2018. A hybrid origin of the Martian crustal dichotomy: Degree-1 convection antipodal to a giant impact. Earth and Planetary Science Letters. Volume 491. Pages 58-66.

109 Clarkson, Euan and Brian Upton. 2009. Death of an Ocean: A Geological Borders Ballad. Dunedin Academic Press.

extensive Iapetus Ocean was closed in the collision of the tectonic plates of Laurentia, Baltica and Avalonia causing the intense folding and uplift of huge quantities of ocean floor to form the Caledonia Mountains. According to others, the western end of the ocean saw the incorporation of an extensive island arc of volcanic islands between the colliding tectonic plates. The transformation by obduction of the Iapetus Ocean to form a mountain range is similar to the formation of the Himalayans. Mantle convection pushed the Indo-Australian plate towards the Eurasian plate and as time went on the Tethys Sea was crushed and thrust over the Eurasian plate forming the Himalayans. Envision the magnitude of forces required to lift the sea floor more than 5 miles above sea level in the instance of Mount Everest. The sea floor, fossils and all, rests on top of the world's highest mountain. Is this the work of the plates moving at 2 centimeters a year? I don't think so.

Volcanic Chains of Fire

Remnants of *chains of fire* are found along the shoreline of the Cambrian landmass of Rodinia. Along the margins of the torn apart continent, relic chains of molten rocks emerged at plate boundaries with the fractured continent. These volcanic island arcs are analogous to today's long chains of volcanic island arcs that circumvent the Pacific Ocean, referred to as the *Ring of Fire*.

Imagine a Cambrian world when thin tectonic plates served as rafts for continents that were sliding over the surface of the Earth while chains of active volcanoes like those of today's volcanoes of Indonesia like Mount Pinatubo, Krakatoa, Mount Tambora, and others release gases to the atmosphere and pour out lava onto Earth's surface. It was a time when the Ross Orogeny took place while deforming and building the Transantarctic Mountains. And it was in this time of continental collisions that sediments appear along the collision zones as extensive siliciclastic and turbidite sequences.[110] But just as this tectonic activity led to the breakup of

---

110 Rowell, A.J. et al. 2001. Latest Neoproterozoic to Mid-Cambrian age for the main deformation phases of the Transantarctic Mountains: new stratigraphic

the Cambrian land form of Rodinia, the continental land masses were themselves undergoing a transformation.

Tsunamis and Mega-Flooding

The Cambrian continental margins and lowlands were virtually continuously flooded with inland seas (epicontinental seas). Throughout the entire Phanerozoic Eon some of Earth's highest sea levels coupled with low elevations of the landscape resulted in the invasion of the seas upon the land. Glaciers, which tie up much of the ocean's water, contributed melt waters to the even higher water levels during the Paleozoic. The Paleozoic was followed by the Mesozoic Era; a period that scientists consider to have lasted about 200 million years. By the end of the Mesozoic Era, the world's oceans were about 600 to 800 feet higher than present, which is thought to be the consequence of the displacement of the sea by the mid-ocean ridges and glacial melt-waters. In the final Quaternary Era, scientists believe that glaciation formed ice that took water back out of the oceans, lowering sea level by ~400 feet below today's levels. One might say that since the dawn of the Cambrian Explosion of life, the land masses of planet Earth, with few exceptions, have been repeatedly under water and ice.

Throughout the Cambrian, indications of widespread flooding and tsunamis are evident from Antarctica, to Australia, Wyoming, Quebec, and Scotland. Given the high elevation of the oceans and the low lying topography of the land masses, a large asteroid impact in oceanic areas would have triggered tsunamis and turbidites that would change the landscape. For example, early Cambrian clastic marine sedimentary rocks are thought to have been rapidly deposited in the Kanmantoo Trough of South Australia. Haines and others[111] argue that:

---

and isotopic constraints from the Pensacola Mountains, Antarctica. Journal of the Geological Society, London. Vol. 158, pp. 295-308.
111 Haines, Peter & Jago, James & Gum, Justin. (2001). Turbidite deposition in the Cambrian Kanmantoo Group, South Australia. Australian Journal of Earth Sciences. 48. 465 - 478.

.. these sandstone beds could be the products of sustained high-density turbidity currents. Triggering mechanisms for such turbidity currents remain uncertain, but they may have been initiated as hyperpycnal flows during catastrophic flood events at the mouths of high- load-capacity rivers, or from the failure of unstable buildups of sediment on delta slopes.

Similarly, Brian Pratt of the University of Saskatchewan concludes that tsunamis played an important role during the Cambrian of Argentina[112] and Montana.[113] Evidence of high capacity turbidites and catastrophic flooding are recorded in the rocks. The Cambrian was a time of elevated temperatures, elevated sea levels, and the break-up and reconsolidation of land masses. It was a time of the formation of new oceans with extensive shorelines and shelves where the first complex sea creatures called home. It was also a time of fire; fire that breathed from below, swelling and raising the crust and erupting its fury on the land. The land masses were dominated by quartz rich granites which floated on the thin ocean crust like rafts upon the sea. And these quartz rich batholiths were eroded again and again by the invading seas, producing enormous quantities of sand that made its way to trenches and seashores. I have come to know this sand which is found in the peaceful little town of Addy just north of Spokane, Washington. The sandstone, shales, and limestone rest on the erosion surface at the base of the great Cambrian Sauk transgression marking the period when the ocean rose and flooded the continent.[114]

---

112 Pratt, Brian R. 2007. Tsunamis in a Stormy Sea: Middle Cambrian Inner Shelf Limestones of Western Argentina, Journal of Sedimentary Research, Volume 77(4). Pages 256-262.
113 Pratt, Brian R. 2002. Storms versus tsunamis: Dynamic interplay of sedimentary, diagenetic, and tectonic processes in the Cambrian of Montana. Geology. Volume 30(5). Pages 423-.
114 Karlstrom, K. and others. 2018. Cambrian Sauk transgression in the Grand Canyon region redefined by detrital zircons. Nature Geoscience.

5 *Geodynamo, Poles, and Ocean Floor Graveyards*

He established the earth upon its foundations, So that it will not totter forever and ever....[115] Raging waves of the sea, foaming out their own shame; wandering stars, to whom is reserved the blackness of darkness forever.[116]

Over the past several decades, scientists have become concerned with the rapid increase in the migration of the Earth's magnetic pole. In 1900, the magnetic North Pole was in Canada. In 2000 it was in Greenland. And now it is racing towards Siberia at a whopping 40 kilometers a year.[117] This is likely due to the behavior of the molten ocean at the core of the Earth. As you probably suspect, the sloshing of this molten ocean around the Earth's solid core can cause instabilities analogous (in a simple way) to the poor alignment of tires on a car. Consequently, it imparts a wobble to the trace of the magnetic field at the Earth's surface. The surface of this deep molten ocean that we refer to as the outer core, is overlain by the Earth's mantle. The interface between the core and mantle is irregular, fostering cross boundary exchanges of heat and matter and associated coupled thermal-mechanical-chemical-plastic flow-and possible nuclear reactions:

The partial collapse of topographic structure at the core-mantle boundary (CMB) in avalanches, slumps or turbidity flows, would cause sudden temperature changes in both the upper core and the lower mantle. Although such collapses

---

115 Psalm 104:5.
116 Jude 1:13.
117 Michael Brooks. June 26, 2019. The North Pole is Moving and if it Flips, Life on Earth is in Trouble. New Scientist.
newscientist.com/article/mg24232360-700-the-north-pole-is-moving-and-if-it-flips-life-on-earth-is-in-trouble/

are hypothetical, it is interesting to investigate the potential consequences. Downwelling from such events could disrupt core convection cells and trigger geomagnetic excursions and reversals. Buoyant sediment from the freezing of the inner core is hypothesized to rebuild the avalanched structures. Large avalanches could trigger Mantle plumes. Oblique extraterrestrial impacts impart high shear to the CMB [core mantle boundary], and can trigger one or more simultaneous avalanches, yielding observed coincidences between craters, tektite fields and reversals. [118]

Circulating Ocean of Molten Metal

Underneath your feet, about 2900 kilometers below the Earth's surface, there is a violent swirling turbulent ocean. This ocean itself is about 2270 kilometers deep, which is deeper than the diameter of the Moon! But it is not the kind of ocean we are accustomed to at the Earth's surface. This deep ocean is made of a molten-metal, iron-rich fluid that swirls above the inner core of the Earth with temperatures on the order of 4500 to 5500 degrees Celsius. This ocean is known as the Earth's outer core. *Apparently*, the inner and outer cores spin ever so slowly in opposite directions.[119] The solid inner core rotates around the same axis and in the same easterly direction as the Earth. Meanwhile, the molten metal ocean is sheared as it spins westwards between the inner core and the mantle. *The magnetic field pushes eastwards on the inner core, causing it to spin faster than the Earth, but it also pushes in the opposite direction in the liquid core, which creates a westward motion.* [120] The seismic boundary between the outer and inner cores is sometimes referred

---

[118] Muller, Richard, A. 2002. Avalanches at the core-mantle boundary. Geophysical Research Letters. Volume 29. Pp. 41-1 to 41-4.
[119] Livermore, P.W., R. Hollerbach, and A. Jackson. 2013. Electromagnetically driven westward drift and inner-core superrotation of Earth's core. PNAS. 110 (40) 15914-15918.
[120] Zolfagharifard, E. 2013. A 300-year-old riddle finally solved: Earth's inner core spins in an eastward direction - the opposite to the outer core. dailymail.com

to as the **Lehmann-Bullen discontinuity**. At this interface, temperatures (about 6000 degrees Celsius) rival the temperatures at the surface of the Sun (about 5527 degrees Celsius).

Eddy currents of swirling iron in the outer core exist in an ionically charged state in the thermally convective fluid which induces the Earth's magnetic field to the tune of about a billion amps! This imparts rotational forces upon the inner core. The spiral rotation of flow caused by the Coriolis effect plays a major role in aligning the flow field. And in turn, the charged ions and free electrons passing through the magnetic field impart electric currents, establishing a self-sustaining loop known as the geodynamo. However, the system is not a **perpetual motion machine**. Crystallization of iron ions within the liquid core transfers heat to the mantle of the Earth and results in growth of the inner core at about a millimeter a year.[121]

Earth's Magnetic Field almost Collapsed

Earth's geodynamo almost died just prior to the Cambrian period. It was at a time associated with the explosion of life on the planet. It was the time of worldwide flooding. It was the time of the break up of Rodinia. It was a time of solid core formation. It was a time when the Earth's magnetic poles were rapidly wandering and flipping. It was a time when oxygen in the atmosphere suddenly increased. Was it a coincidence, or are all of these factors somehow related? I have a hunch they're all related. How about you? Without the Earth's geodynamo there would be no means of protecting life from the deadly solar winds which bombard the planet with charged particles referred to as cosmic radiation. But what caused the geodynamo to weaken and what re-started the Earth's geodynamo? The Earth's magnetic field is recorded by nature in rock formations. As magma cools, the orientation of magnetic minerals like magnetite (Fe_3O_4) align with the Earth's paleo-magnetic field. A team of scientists from the University of

---

121 Waszek, Lauren; Irving, Jessica; Deuss, Arwen (2011). "Reconciling the hemispherical structure of Earth's inner core with its super-rotation". Nature Geoscience. 4 (4): 264–267.

Rochester[122] took rock samples from Quebec, Canada and found that these rocks recorded an interval in Earth's history *in the Cambrian when the Earth's magnetic field was a tenth of today's intensity.* As the solid core of the Earth spins, the ocean of liquid iron that makes up the outer core spins more slowly.

The liquid iron in the outer core of Earth moves by convection causing magnetic lines to twist, thereby continuously generating the magnetic field. According to the researchers, a big boost was needed to keep the field from collapsing altogether. According to one of the researchers, [123] *If the geodynamo had collapsed entirely, Earth would not have been protected from the solar wind, which can erode the atmosphere and eventually rob water from the planet.* Researchers at the University of Rochester suggest that the geodynamo got a kickstart: *iron began to cool and freeze into a solid layer in the middle of the planet. As the inner core solidified, lighter elements like silicon, magnesium and oxygen were kicked out into the outer, liquid layer of the core, creating a movement of fluid and heat called convection. This movement of fluid in the outer core kept charged particles moving, creating an electrical current, which in turn created a magnetic field.*

The weakened state of the Earth's geodynamo in the Cambrian could well have been due to the impacts of one or more asteroids. And later, the geodynamo could have recovered. Perhaps the precipitation and accumulation of iron which led to the Earth's solid core was in some way influenced by such an asteroid impact or impacts as described above by Muller. According to W. Brian Tonks and H. Jay Melosh,[124] *once a planet reaches a certain minimum mass, large impacts characteristic of late accretion can trigger core formation.*

---

122 Osborne, Hannah. 2019. Earth's Magnetic Field was on the Brink of Collapse 565 Million Years Ago. Newsweek.
123 Sapalakoglu, Yasemin, 2019. Earth's Magnetic Field Nearly Disappeared 565 Million Years Ago. www.livescience.com/64625-earth-magnetic-field-nearly-disappeared.html
124 W. Brian Tonks and H. Jay Melosh. 1992. Core Formation by Giant Impacts. Icarus. Volume 100, Issue 2, pages 326-346. Elsevier.

Magnetic Field of Mars Collapsed

Apparently, Mars had a magnetic field like that of Earth's.[125] *A model of asteroids striking the red planet suggests that, while no single impact would have short-circuited the dynamo that powered its magnetism, a quick succession of 20 asteroid strikes could have done the job. "Each one crippled a little bit," said geophysicist Jafar Arkani-Hamed of the University of Toronto, author of the new study. "We believe those were enough to cripple, cripple, cripple, cripple until it killed all of the dynamo forever."* According to Arkani-Hamed (in the same article) a mega impact of a Texas size asteroid hit Mars shutting down the convection of the dynamo. ***But left alone, convection would have recovered** in the outer parts of the core, and eventually penetrated deep and started the whole core churning again. The Borealis impact would have crippled the dynamo, but not killed it outright.* Perhaps in similar fashion it was the bombardment of asteroids that weakened the Earth's own geodynamo at the beginning of the Cambrian, followed by rapid reversals[126] signaling a weakened field. Could the restart of Earth's magnetic field in the Cambrian be partly associated with the developing crustal graveyards at the core mantle boundary?

Graveyards of Broken Ocean Floors

Legends of lost cities drowned by the sea have long stirred the imaginations of men. The lost city of Atlantis has been the subject of deep sea expeditions. In about 360 BC, the Greek philosopher Plato, described a utopian society of half mortals and half gods. Plato tells us that the God Poseidon named his son Atlantis, the eldest of ten brothers, as the heir named to govern the kingdom.

---

125 Crossman, Lisa. 2011. Multiple Asteroid Strike May Have Killed Mars's Magnetic Field. www.wired.com/2011/01/mars-dynamo-death/
126 Zongqi Duan, Qingsong Liu, Shoumai Ren, Lihui Li, Xiaolong Deng, Jianxing Liu, Magnetic reversal frequency in the Lower Cambrian Niutitang Formation, Hunan Province, South China, Geophysical Journal International, Volume 214, Issue 2, August 2018, Pages 1301–1312

As shown in the map below, the kingdom was located somewhere between America and Africa (note that the map is "upside" down).

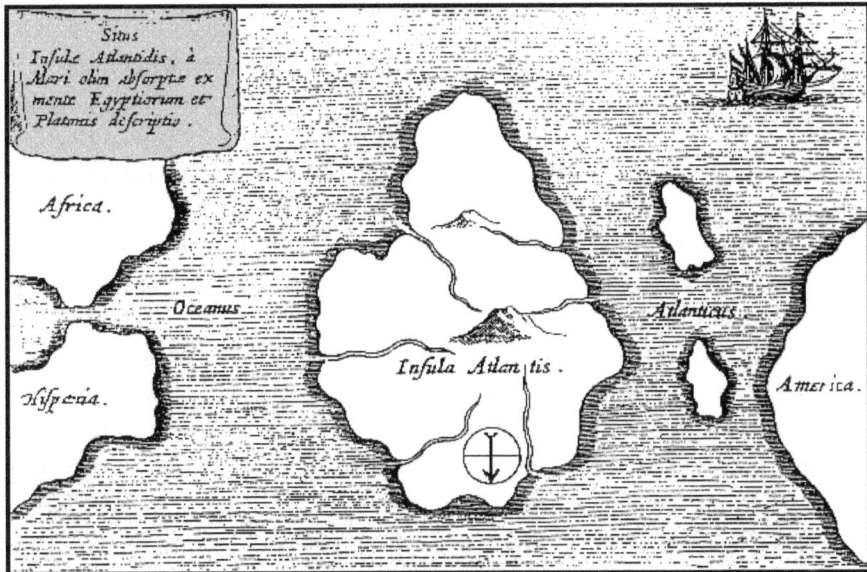

Athanasius Kircher's Atlantis circa 1680[127] Note North is at the bottom of the sketch and South is at the top.

Islands like Krakatoa have been lost below the surface of the sea; a possible fate like the fabled lost land of Atlantis. Recently, scientists claim that they've found a lost continent in the Pacific Ocean. It covers an area of 4.5 million square kilometers. And it only rises above the Pacific Ocean surface in New Zealand and New Caledonia.[128] Drowned cities have been found in many places around the globe but what about lost oceans?

Entire non-fictional oceans have disappeared from the surface of the earth. New ocean floor is generated along the spine of the mid-Atlantic Ridge and other spreading centers. In the North Atlantic, the extension of the mid-Atlantic Ridge projects through Iceland's Thingvellir Rift Valley. Along the rift valley one can

127 Kircher, Athanasius. 1680. Public Domain ⌖ download from Wikimedia commons. July 2019.
128 Gosnell, Peter. 2017, Scientists claim existence of drowned Pacific Ocean continent. Livemint.com.

walk along the base of a high wall on one side of the valley where the North American plate resides and then take a short walk across the narrow valley to the wall on the other side where the Eurasian plate resides. The rift zone is a center of seismic activity where mantle flow is leading to the slow separation of the two walls. If the activity continues, it will provide an opening for the Atlantic ocean to flow through. This early stage of rifting is analogous to the early stages of the opening of the Gulf of Adan and the Red Sea in the Middle East and the extension of their rifts at the triple junction of the continental breakup of Africa referred to as the Great African Rift Valley. *But where did these older oceans go?*

In the Cambrian, the lost ocean that geologists call Iapetus preceded the opening of the early Atlantic. But what was the fate of Iapetus? Geologists conclude that as plates moved upon the surface of the upper mantle, Iapetus was over-ridden by colliding plates leading to its demise. Likewise, the Tethys Sea met a similar end. But where did its oceanic plates go? We know that much of the sea floor was thrust upwards in the building of the Himalayas. Doctors use ultrasounds to examine patients. Perhaps you've watched an ultrasound. Similarly, geologists use seismic tomography to examine the internal structures of Planet Earth. Through these studies, geologists have found an enormous graveyard of fragments of these lost oceans resting at the bottom of the mantle right upon the Earth's outer core. Two major provinces have been discovered which have large accumulations of drowned ocean plates: one beneath the Atlantic ocean and Africa, the second beneath the Pacific ocean.[129] The following sketch is for conceptual purposes and is adapted from technical papers on the subject.[130] The purpose is to illustrate the enormity of the

---

129 Paula Koelemeijer, Arwen Deuss & Jeroen Ritsema. 2017. Density structure of Earth's lowermost mantle from Stoneley mode splitting observations
Nature Communications volume 8, Article number: 15241
130 See a review of numerous papers by Allen K. McNamara. 2019 (Article in Press). A review of large low shear velocity provinces and ultra-low velocity zones. Tectonophysics. www.elsevier.com/locate/tecto see also

"thermochemical piles"[131] or subducted ocean floor resting on the mantle. Note that mantle plumes typically form at the edge of these piles. As new seafloor is generated at spreading centers, seafloor is driven under continents and sinks deep within the mantle, and collects enmasse like shipwrecks at the Mantle-Core boundary. Are these graveyards a part of recycle with the Crust?

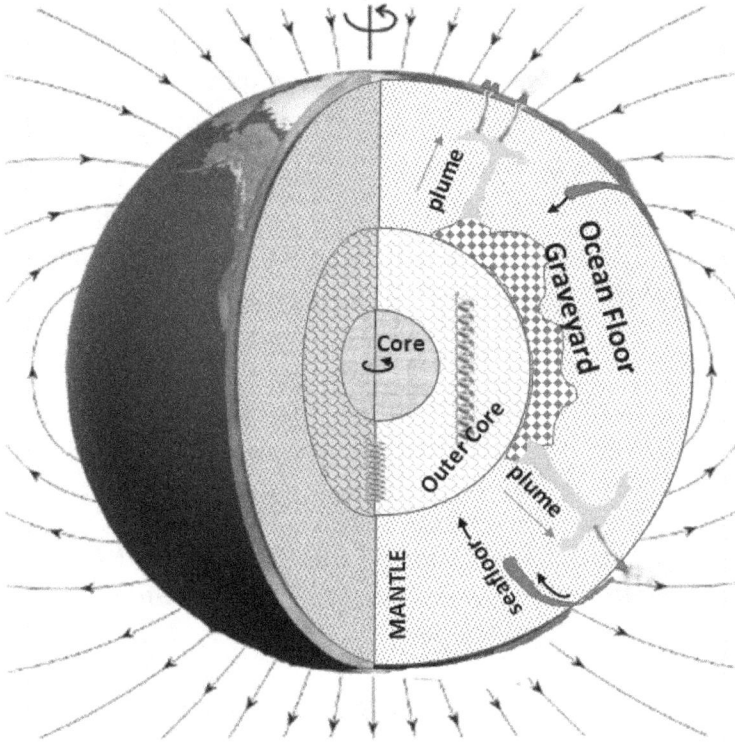

Could temperatures buildup below these graveyards at the interface between the Core and Mantle? Would elevated temperatures that are evidenced by the release of superplumes from their margin be sufficient to encourage nuclear reactions? Maybe!

A. L. Bull, A. K. McNamara, and J. Ritsema, "Synthetic tomography of plume clusters and thermochemical piles," Earth and Planetary Science Letters, vol. 278, no. 3-4, pp. 152–162, 2009.
131 Powell, Corey S. 2019. Deep inside Earth, scientists find weird blobs and mountains taller than Mount Everest. nbcnews.com/mach/science

6 Fountains of the Great Deep

For the invisible things of him from the creation of the world are clearly seen, being understood by the things that are made, ... so that they are without excuse:[132] the same day were the fountains of the great deep broken up, and the windows of heaven were opened. And the rain was upon the earth forty days and forty nights. ....[133] he established the clouds above: when he strengthened the fountains of the deep...[134]

Scientists scoff at a world-wide Flood. I know I did. Afterall, where is the evidence? If there was such a world-wide Flood, where did all the water go? And we are told that the water came from below and above. Really? *Is there any evidence that the salt waters of our great oceans once resided below the surface of our planet? And what are these Fountains of the Great Deep?* The Scriptures say the evidence is clearly seen by things that are made. For most of my life, I thought these Scriptural statements were beyond the reasonableness of physical reality. Strangely enough, astronomers have been looking at some of the evidence for hundreds of years. Recently, a web article titled, *"Mars' Chicxulub Crater" –Monster Tsunami Left Evidence of a Vast Ocean (& Ancient Life?)* was published.[135]

---

132 Romans 1:20
133 Genesis 7:11, 12.
134 Proverbs 8:28.
135 The Daily Galaxy via BBC, Nature, New York Times Science, NASA Ames Research Center. Posted on-line July 31, 2019. *"Mars' Chicxulub Crater" –Monster Tsunami Left Evidence of a Vast Ocean (& Ancient Life?).* dailygalaxy.com

According to the article, Mars had an ancient ocean on its northern hemisphere. A team of French scientists claim that a huge asteroid slammed into Mars and caused an initial mega-tsunami wave estimated at 300 meters high (nearly 1000 feet high). They postulate that a second giant wave raced across the Planet at 120 meters high (nearly 400 feet high), traveling at an estimated 200 feet per second. More importantly, the report concludes:

> *The ocean may have been fed by catastrophic floods from <u>underground caches of liquid water</u>. If so, sediments in the north "may be a window into the subsurface habitability of Mars," Rodriguez said. If they contain geochemical signatures of ancient microbiology, then the liquid aquifers thought to exist beneath Mars' surface may still be reservoirs for life today.*

Doesn't this sound familiar? These results are comparable to the modeling of the tsunamis generated by the Chicxulub impact! Maybe evidence of catastrophic processes will clearly be seen elsewhere in our Solar System.

January 7, 1610 Letter on the Discovery of the 4 Galilean Moons

Astronomer Galileo Galilei aimed his telescope upon Jupiter and soon identified four small spherical bodies orbiting about the giant planet. Galileo documented the four moons in a letter, dated January 7, 1610. Although Galileo named the moons, a contemporary German astronomer, Simon Marius of Ansbach gave the four moons formal names: Io, Europa, Ganymede and Callisto. More importantly, it is through the study of these moons and a number of other moons and planets of our Solar System, that important light is shed upon processes that may have been active right here on Planet Earth. And this extraterrestrial evidence may help answer the riddle of the *Fountains of the Great Deep!* Processes observable on two moons within our Solar System, Ganymede and Enceladus are particularly relevant.

Evidence from Ganymede: Sloshing Subsurface Salt Ocean Impacts Moon's Magnetic Field

Ganymede, one of Jupiter's moons discovered by Galileo is the largest known moon in our Solar System and the only moon known to have a magnetic field. Interestingly enough, it was the Galileo mission that found the moon's subsurface ocean and it's magnetic field. Ganymede is larger than our moon and a little less than half the radius of Earth but larger than the planet Mercury. Ganymede has an iron-rich liquid core like that on Earth and a subsurface ocean that is thought to have more water than all of Earth's oceans. According to Kivelson et al:[136]

> *An inductive response is consistent with a buried conducting shell, probably liquid water with dissolved electrolytes, somewhere in the first few hundred km below Ganymede's surface. The depth at which the ocean is buried beneath the surface is somewhat uncertain, but our favored model suggests a depth of the order of 150 km...*

So the circulating subsurface ocean acts as a conductor which is postulated to cause disturbances to Ganymede's magnetic field. *Could the sloshing of a past surface ocean or even a subsurface ocean have affected Earth's magnetic field? I think so. How about you?*

Evidence from Enceladus: Fountains of the Great Deep

The Cassini Mission found that Saturn's moon Enceladus spews water hundreds of kilometers into space. More than a hundred such fountains are found venting from four 84 mile long parallel fractures near the moon's south pole.

---

136 Kivelson, M.G., K.K. Khurana, and M. Volwerk. 2001. The Permanent and Inductive Magnetic Moments of Ganymede. Icarus 157. 507-522.

APOCALYPSE NOW: THE ROCKS CRY OUT

During a recent stargazing session, NASA's Cassini spacecraft watched a bright star pass behind the plume of gas and dust that spews from Saturn's icy moon Enceladus. At first, the data from that observation had scientists scratching their heads. What they saw didn't fit their predictions. The observation has led to a surprising new clue about the remarkable geologic activity on Enceladus: It appears that at least some of the narrow jets that erupt from the moon's surface blast with increased fury when the moon is farther from Saturn in its orbit.[137]

According to NASA, *Cassini revealed the dramatic truth: Enceladus is an active moon that hides a global ocean of liquid salty water beneath its crust. What's more, jets of icy particles from that ocean, laced with a brew of water and simple organic chemicals, gush out into space continuously from this fascinating ocean world. The material shoots out at about 800 miles per hour (400 meters per second) and forms a plume that extends hundreds of miles into space. Some of the material falls back onto Enceladus, and some escapes to form Saturn's vast E ring.*

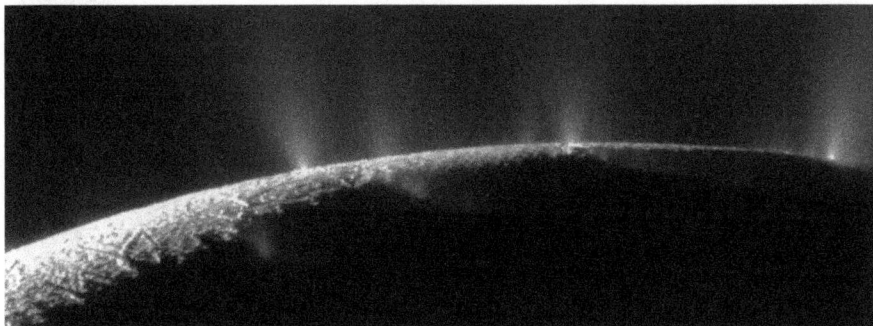

Image Credits: NASA/JPL/Space Science Institute. Caltech.

---

137 NASA. May 2016. Enceladus Jets: Surprises in Starlight. NASA.gov
See also: New Study Says Enceladus has had an Internal Ocean for Billions of
Years. universetoday.com

Sounds like a Snowball Earth! The geysers are now known to have their origin in an ocean 16 to 19 miles below the surface of Enceladus. Importantly, the surface of Enceladus is minus 330 degrees Fahrenheit, whereas samples of the jet spray include ice crystals, gases, and silica suggesting that the jets come from deep hydrothermal vents.

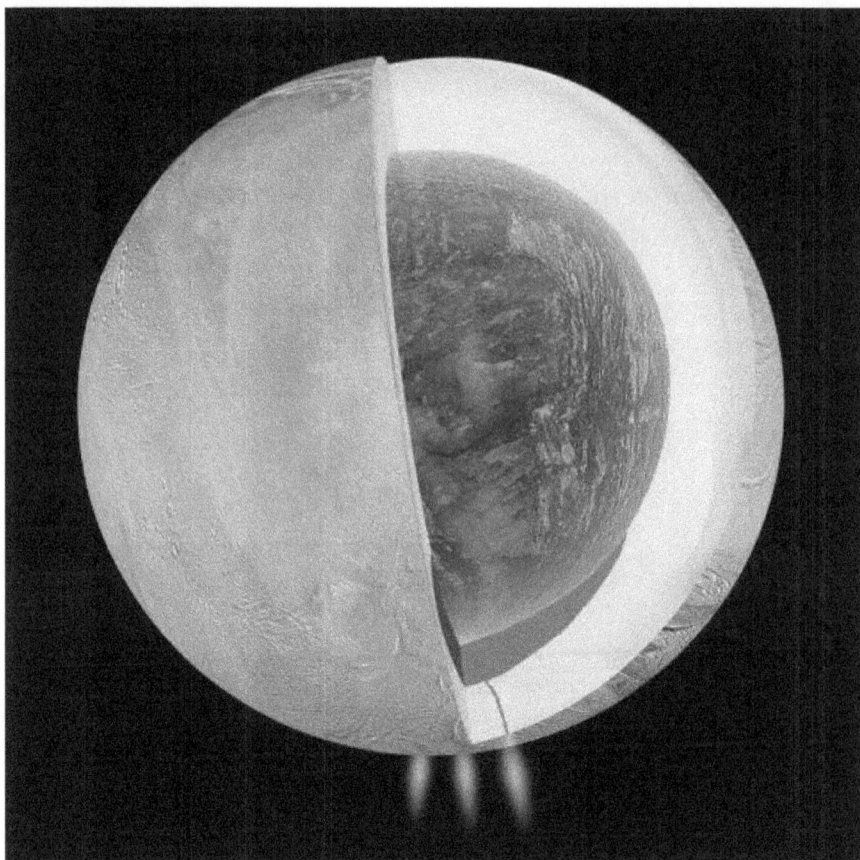

Several gases, including water vapor, carbon dioxide, methane, perhaps a little ammonia and either carbon monoxide or nitrogen gas make up the gaseous envelope of the plume, along with salts and silica. And the density of organic materials in the plume was about 20 times denser than scientists expected. Since the ocean in Enceladus supplies the jets, and the jets produce Saturn's E ring, to

study material in the E ring is to study Enceladus' ocean. The E ring is mostly made of ice droplets, but among them are peculiar nanograins of silica, which can only be generated where liquid water and rock interact at temperatures above about 200 degrees Fahrenheit (90 degrees Celsius). This, among other evidence, points to hydrothermal vents deep beneath Enceladus' icy shell, not unlike the hydrothermal vents that dot Earth's ocean floor.[138]

If the jet process is occurring on a moon of Saturn, I suspect similar processes could have happened on early Earth, after all, some refer to the early Planet Earth as *Snowball Earth*.

When the Snowball events occurred, the supercontinent Rodinia was in the process of breaking up. A supercontinent is a state in which all of the continents are clustered together in one group.[139]

Could analogous processes on Earth have caused jets that we call the *Fountains of the Great Deep*? Could large quantities of water have been stored at depths and pressures comparable to the subsurface ocean of Enceladus? Could similar processes have been associated with *Snowball Earth* and the breakup of Rodinia? I believe so. How about you?

---

138 NASA. April 2019. Enceladus In Depth; By the Numbers. NASA.gov

139 Poppick, Laura. April 2019. Snowball Earth: The times our planet was covered with ice. Astronomy.com See also NASA. "Snowball Earth" Might Have Been Slushy *By Michael Schirber, Astrobiology Magazine-August 2015.*

7   Diamonds: Windows to the Mantle's Hidden Oceans

...the same day were all the fountains of the great deep broken up, and the windows of heaven were opened.[140] But where shall wisdom be found? and where [is] the place of understanding? ...The depth saith, It [is] not in me: and the sea saith, [It is] not with me. It cannot be gotten for gold, neither shall silver be weighed [for] the price thereof. It cannot be valued with the gold of Ophir, with the precious onyx, or the sapphire. The gold and the crystal cannot equal it: and the exchange of it [shall not be for] jewels of fine gold. Whence then cometh wisdom? and where [is] the place of understanding?[141]

Could the Earth be hiding the source of the Scriptural *Fountains of the Great Deep?* Does the Earth's mantle hold oceans of water like those on Enceladus and Ganymede? Diamonds carry a highly pressurized cargo from great depths of the mantle to the surface. Through them we can learn about ancient oceans, the composition of the mantle, and the origin of the Earth's crust. For it is *within these minerals that we find small capsules, called fluid inclusions, that provide a window into the Great Deep,* where mighty fountains arise. These samples are brought to the surface by kimberlite, lamproite, and carbonatite magmas, and are our means of understanding the chemistry of the mantle. Wisdom is prized more than diamonds yet we study minerals to find wisdom. We use them to establish temperatures and pressures from deep within the Earth. Fluid inclusions act as a deep sampling device that encapsulate liquids, gases, and solids that were present when they were growing. You might say these inclusions provide libraries of information about the Earth's deep crust and mantle.

---

140 Genesis 7:11.
141 Job 28:12 – 20.

The Race to Reach the Earth's Mantle

When I was a boy, I was fascinated by two races in the sciences: the race to drill a deep hole into the Earth to sample the mantle. The second was a race into space to take a sample from the surface of the moon. The first of the two initiatives was given birth in 1957. Both would provide strategic advantages in the exploration and exploitation of resources for economic and political survival.

The ***Race to Reach the Earth's Mantle*** was proposed during a 1957 session of the National Academy of Sciences (NAS). While the NAS was reviewing Earth Science proposals it became captivated with the proposal by Dr. Walter Munk of Scripps Institution of Oceanography. The proposal received support from Dr. Harry Hess, a professor at Princeton, an early proponent of plate tectonics, and mentor to one of my college professors. The initiative took on a competitive political twist when the Russians launched Sputnik and were later rumored to be conducting a drilling program of their own. The early efforts to drill to the boundary between the Earth's crust and mantle, referred to as the Mohorovicic Discontinuity focused on drilling where Earth's crust was assumed to be thinnest. Drilling began in earnest off the coast of Guadalupe, Mexico. The boring penetrated through 557 feet of young crustal sediments. Once through the sediments, a second layer of basalt crust was penetrated. The deepest boring went to 601 feet below the sea floor which was at a depth of 11,700 feet.[142] The Russians ultimately continued on their Kola Superdeep Borehole, reaching a depth of 7.6 miles.[143]

But as deep as these boreholes are, they barely scratch the surface. The crust of ocean floor is about 5 to 10 kilometers thick and the thickness of continental crust varies between 20 to 90 kilometers. However, earth scientists have been working around these limitations. First, as nature would have it, there are numerous well studied areas where the Earth's mantle has been brought to the surface in what are referred to as ophiolite complexes. An

---

142 See details in en.wikipedia.org/wiki/Project_Mohole
143 See details in en.wikipedia.org/wiki/Kola_Superdeep_Borehole

ophiolite complex is a segment of ocean crust and upper mantle tectonically lifted up from great depth along boundaries of the Earth's surface where the giant plates push against each other. These are typically found in areas where mantle rock below the ocean floor is pushed up over continents. Many localities of these complexes are associated with the closure of the Tethys Ocean where the uplifted rock sequence is overlain by turbidites.[144] Likewise, remnant ophiolite complexes from the closure of the Iapetus Ocean are observable on the surface, in places where they have been thrust up over the continents that once fenced them in.

> *About 480 million years ago the Iapetus Ocean began to close and by 420 million years ago the continents on either side - early versions of North America, Europe and Scandinavia - had collided to form the continent of Euramerica (uniting the rocks of Scotland and England for the first time).[145]*

Interestingly, there are three major periods of ophiolite complexes being thrust over continental crust: Precambrian, Cambrian, and Triassic/Jurassic/Cretaceous and these three periods are associated with three worldwide magmatic events. They are found associated with subduction zones and volcanic island chains (island arcs are volcanic islands arranged like a string of pearls like those of Samoa and the Philippines). Some are found on the San Juan Islands of Washington. But even these deposits barely scratch the surface. But they do provide an examination of the rocks that were once at the Moho![146] Sound waves travel 343 meters/second in air; and they travel at 1,480 meters/second in water; and even

---

144 Zhen Yan and others. 2019. Early Cambrian Muli arc-ophiolite complex: a relic of the Proto-Tethys oceanic lithosphere in the Qilian Orogen, NW China. International Journal of Earth Sciences, Volume 108, Issue 4,; pages 1147-1164.
145 www.shetlandamenity.org/shetlands-ophiolite. See also Flinn, Derek, P. Stone, and D. Stephenson. 2013. The Dalradian rocks of the Shetland Islands, Scotland. Proceedings of the Geologist's Association. Vol. 124, Issues 1-2. Pages 393-409.
146 Garver, John, I. 2011. Stratigraphy, depositional setting, and tectonic significance of the clastic cover to the Fidalgo Ophiolite, San Juan Islands, Washington. Canadian Journal of Earth Sciences. Volumen 25(3):417-432

faster in iron at 5,120 m/s (15 times as fast as in air). The Mohovorovicic discontinuity (Moho) is defined as that level in the Earth where the compressional wave velocity increases rapidly or discontinuously; that is, by a distinct change in the velocity of seismic waves. Above the Moho boundary, we find basalts with seismic velocities of 6.7 to 7.2 km/second. Below the Moho boundary we find a more dense suite of rocks known as peridotites and dunites (dense, coarse-grained rocks containing large amounts of olivine)[147] with a jump in wave velocities to 7.6 and 8.6 km/second.[148] Based on these studies we have great confidence in defining even deeper discontinuities (zones where seismic velocities change). Significant discontinuities (marked by changes in seismic velocities) are known to occur at the mantle-core boundary and the outer core-inner core boundary.

Superplumes and Samples from the Deep Mantle

Earlier we discussed the graveyards of submerged ocean floor slabs and their fragments resting at the boundary between the Earth's mantle and core. These graveyards apparently include massive numbers of slabs from the Iapetus and Tethys oceans. Seismologists refer to these graveyards as thermo-chemical piles. They are also referred to as large low shear velocity provinces (LLSVPs) because of the seismic properties of the large volumes of these buried crustal slabs. Importantly, they exhibit sharp seismic velocity gradients at their margins.[149] The deep pile-up of ocean-floor crust in these graveyards play numerous roles with respect to the habitability of our planet. In one sense they "seal-off" and insulate upwelling currents and magma from the core and

---

147 Olivine is a mineral of iron, magnesium, silicon, and oxygen with the formula (Fe, Mg)2SiO4

148 Jarchow, Craig M. and George A Thompson. 1989. The Nature of the Mohorovicic Discontinuity. Annual Review of Earth and Planetary Sciences, Vol. 17, page 475-505.

149 Davies, D. Rhodri and S. Goes. 2015. Thermally dominated deep mantle LLSVPs: a Review. Book Chapter 14, pages 441-477, in: The Earth's Heterogeneous Mantle: A Geophysical, Geodynamical, and Geochemical Perspective. Springer Geophysics. ISBN 978-3-319=15627-9.

divert superplumes, forcing them to rise from their perimeters. *The LLSVPs act as thermal insulators, making the core-to-mantle heat transfer by conduction effectively focused at the edges of the LLSVPs, where thermal diapirs dominated by SML material develop and rise as plumes, explaining why LIPs* [large igneous provinces] *are projected to the LLSVP edges.*[150] In turn, these superplumes of upwelling magma, impact the Earth's magnetic field and may even cause pole reversals.

> *Superplume growth increases the mean core–mantle boundary heat flux and its lateral heterogeneity, thereby stimulating polarity reversals, whereas superplume collapse decreases the mean core–mantle boundary heat flux and its lateral heterogeneity, inhibiting polarity reversals. Our results suggest that the long, stable polarity geomagnetic superchrons such as occurred in the Cretaceous, Permian, and earlier in the geologic record were initiated and terminated by the collapse and growth of lower mantle superplumes, respectively.*[151]

Most importantly, for our discussion, rocks and minerals are brought from the core mantle boundary by these superplumes. Diamonds and fragments of the mantle, are brought to the Earth's surface by kimberlites, lamproites, and carbonatites like those I sampled near McClure Mountain, Colorado. These types of magmas rise to the surface propelled extremely rapidly and violently by gases such as carbon dioxide from great depths where the pressures are as much as 60,000 times that on the Earth's surface.

> *Plate reconstructions and tomographic images … show that the edges of the largest heterogeneities in the deepest mantle,*

---

150 Niu, Yaoling. 2018. Origin of the LLSVPs at the base of the mantle is a consequence of tectonics – A petrologic and geochemical perspective. Geoscience Frontiers 9. Pages 1265-1278.
151 Amit, Hagay and Peter Olsen. 2015. Lower Mantle Superplume growth excites geomagnetic reversals. Earth and Planetary Science Letters. Vol. 414, Pages 68-76.

… seem to have controlled the eruption of most Phanerozoic kimberlites.[152]

Diamond mining at the Kimberley Mine in South Africa is one example of nature's own ability to bring samples of the lowest reaches of the mantle to the surface for man's investigation. Interestingly, scientists have found evidence of a substance previously thought only to exist on extraterrestrials like Enceladus.

Diamonds Reveal a Mega-Ocean Far Below Us

Numerous headlines are announcing a revolution in the earth sciences: *The World's Deepest, Rarest Diamonds Revealed a Big Secret About Our Planet's Interior*[153] and *Huge Amount of Water in Earth's Mantle*[154] are leading to *The Hunt for Earth's Deep Hidden Oceans.*[155] Why is the media making such a fuss over an ocean beneath our feet? Researchers are reporting that they are finding a form of water deep below us in the Earth's mantle: a form previously thought only to exist on extraterrestrial planets and moons like Enceladus. Scientists are finding a rare form of water referred to as *Ice VII in diamonds* from deep within the Earth. Some are reporting that this *weird water phase "ice-VII" can grow as fast as 1,000 miles per hour*[156] reporting that this *exotic form of ice could freeze an alien ocean in a few hours*. What does this all mean? Seismologists have defined a *"transition zone"* that resides between the Earth's upper and lower mantle. Some suspect *pockets of water may lie deep below Earth's*

---

152 Torsvik, T.H. and others. July 2010. Diamonds sampled by plumes from the core-mantle boundary. Nature 466. Pages 352-355.
153 Specktor, Brandon. 2018. The World's Deepest, Rarest Diamonds Revealed a Big Secret About Our Planet's Interior. LiveScience. www.livescience.com
154 Renaissance Universal. May 4, 2014. Huge Amount of Water in Earth's Mantle. sureshemre.wordpress.com
155 Woo, Marcus. 2018. The Hunt for Earth's Deep Hidden Oceans. Quantamagazine, quantamagazine.org
156 Ouellette, Jennifer. 2018. Weird water phase "ice-VII" can grow as fast as 1,000 miles per hour. arstechnica.com/science.

surface[157] within and just below this ***transition zone***. Seismic studies indicate that the ***transition zone*** ranges from ~410 to 660 kilometers below the Earth's surface. This ocean is not a body of free flowing water. Rather, this ocean is really better viewed as a reservoir of water. In this reservoir, water is associated with a plastic mush of minerals and ions; all under tremendous pressure.

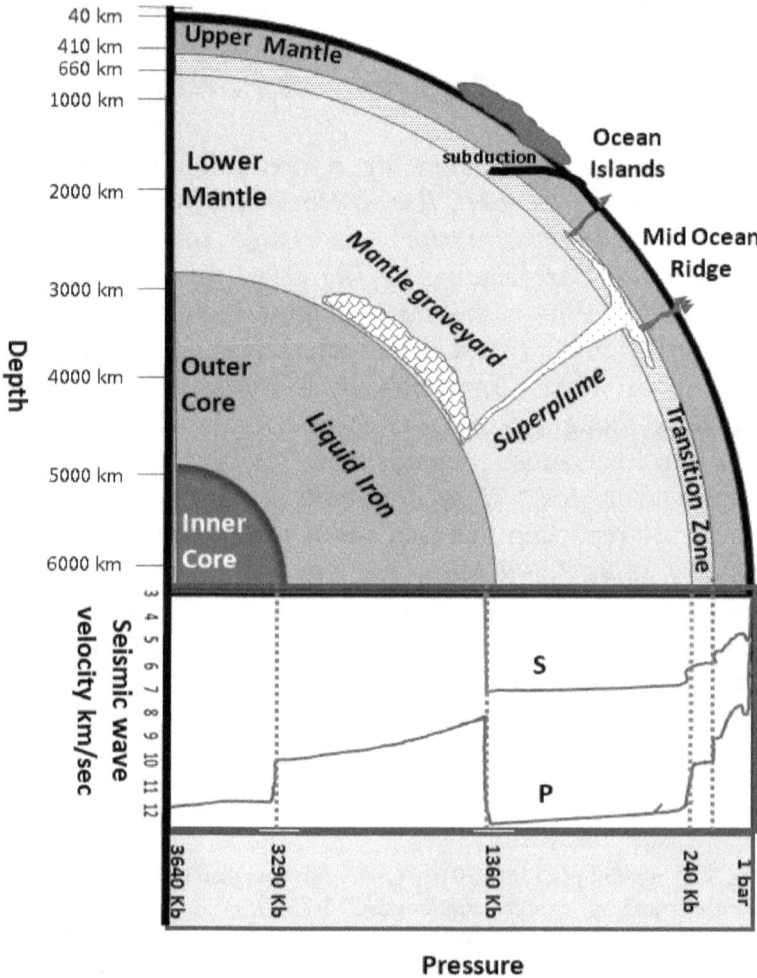

---

157 Perkins, Sid. 2018. Pockets of water may lie deep below Earth's surface. Science Magazine. www.sciencemag.org

Ice VII: Proof of Deep Ocean Reservoirs

A tiny impurity of ringwoodite (a compressed form of olivine) just a few millimeters long, found in a Brazilian diamond, leads scientists to believe that there are significant amounts of Mantle water. The title of the Scientific American article reads, *Rare Diamond Confirms That Earth's Mantle Holds an Ocean's worth of Water*.[158] Ringwoodite only forms under extreme pressures, confirming that it came from more than 500 kilometers into the mantle. This is in the middle of the mantle Transition Zone (410 to 660 kilometers below the surface). The water that has everyone excited represents about 1.5 weight percent of ringwoodite which, if available as free water, would yield the equivalent of an ocean of water. This suggests that there is far more water to be found.

But what about the buzz concerning Ice VII? Ice VII is thought to form the ocean floor of Europa and a significant portion of the ice shells of Enceladus. But over the last couple years, Ice VII has been found in diamonds from below the Transition Zone of the Earth's mantle. Ice VII is the second and confirming line of evidence. In this case, the water is free, not bonded as is the case of ringwoodite. Like ringwoodite, the water is so compressed that the ice is stabilized as the "mineral" solid, Ice VII (H_2O). A report in Science[159] provides evidence for this "free water."

> *Water-rich regions in Earth's deeper mantle are suspected to play a key role in the global water budget and the mobility of heat-generating elements. We show that ice-VII occurs as inclusions in natural diamond and serves as an indicator for such water-rich regions. Ice-VII, the residue of aqueous fluid present during growth of diamond, crystallizes upon ascent of the host diamonds but remains at pressures as high*

---

158 Oskin, Becky. 2014. Rare Diamond Confirms That Earth's Mantle Holds an Ocean's Worth of Water. LiveScience. www.scientificamerican.com
159 Tschauner, O. and others. 2018. Ice-VII inclusions in diamonds: Evidence for aqueous fluid in Earth's deep mantle. Science. Vol. 359, Issue 6380, pages 1136-1139.

as 24 gigapascals; it is now recognized as a mineral by the International Mineralogical Association. In particular, ice-VII in diamonds points toward fluid-rich locations in the upper transition zone and around the 660-kilometer boundary.

Could this water rich zone contribute to the deflection of seismograms? The team drew two important conclusions:

(i) The aqueous inclusions were entrapped as <u>fluid rather than solid ice</u>. Crystallization into ice-VII has occurred at much shallower depth during ascent. (ii) Despite marked uncertainties in the equations of state, the entrapment pressures for the ice inclusions that are currently at ~8 to 12 GPa turn out to be sufficiently narrow to permit a statement of the depth of their source regions: They range from 400 to 550 km depth. For the inclusions at 24 to 25 GPa, the source region is less narrowly estimated at 610 to 800 km depth.

For their first conclusion Tschauner and others rely on work by Craig Bina and Alexandra Navrotsky:[160] *the stable phase of H_2O should be solid ice VII in portions of the coldest slabs. The formation of ice VII as a dehydration product would affect the generation, storage, transport and release of water in cold subduction zones and equilibrium conditions of dehydration would shift, potentially affecting the depths of seismogenesis and magmagenesis. Large amounts of pure ice VII might accumulate during subduction and, <u>as a sinking slab warms, eventual melting of the ice would release large amounts of water in a small region over a short period of time, with a significant positive volume change.</u>*

Water from the dewatering of sea floor could be enormous and would expand enormously if carried back to the surface, reacting with minerals, and leading to the doming of the crust.

---

160 Bina, R. Craig and Alexandra Navrotsky. 2000. Nature. 408, pages 844-847.

8 Earth: Miraculously Planted, Shielded for Life

Earth was without form, and void; and darkness [was] upon the face of the deep. [And] in his hand [are] the deep places of the earth:[161] [and with] a refiner's fire...as a refiner and purifier of silver.[162] All they [are] brass, and tin, and iron, and lead, in the midst of the furnace; they are [even] the dross of silver. [As] they gather silver, and brass, and iron, and lead, and tin, into the midst of the furnace, to blow the fire upon it, to melt [it]... God made the firmament, and divided the waters which [were] under the firmament from the waters which [were] above the firmament: and it was so.[163]

From the beginning, it was as if Earth was deliberately planted in the midst of the Interstellar Cloud. It was neither too close, nor too far, from the Sun for life. Mercury, the Solar System's innermost planet, has a surface temperature of about plus 450 degrees centigrade; it has an exceedingly thin atmosphere with an iron core that makes up about 75% of its radius. Conversely, the outermost planet, Neptune, is the coldest planet in the Solar System with surface temperatures of about minus 200 degrees centigrade. It's atmosphere is composed largely of hydrogen and helium with ices of water, ammonia, and methane. It is considered an ice giant because it is comprised of these ices and rock. Yet, like the Earth, Neptune's core reaches temperatures near 7000 degrees centigrade. And *of all the planets, Earth is the only planet known to have surface water bodies. It is*

---

161 Psalm 95:4, 5.
162 Malachi 3:2,3.
163 Genesis 1:2-10.

the only one with an atmosphere to support life. And by its position in the Solar Winds it receives an ideal balance of chemical compounds and the balance of solids, water, and gases needed for life. Was this by accident or Intelligent Design?

In the beginning low density interstellar clouds of gases and dust were contracting by gravity within the whirling dervish of the interstellar clouds. It was a time when the Solar System's dispersed matter was coalescing. And angular momentum drove these swirling gases and dust clouds into circular orbits much like the rings of Saturn. And like the rings of Saturn, the cloud containing our proto Solar System began to collapse into the disk shape that it exhibits today because its rotation was more efficient along the spin axis of the clouds. As time progressed dust coalesced into planetesimals. Collisions among these formless objects proceeded until a discernable Solar System began to take shape. And through growth by collision and accretion, planetesimals grew into planets.

The size and composition of the resulting planets depended upon their distance from the Sun. Planets coalescing furthest from the Sun accumulated large quantities of frozen ammonia, methane, and water. Planets coalescing nearest the Sun were accreting solid compounds of metals while gases were lost to space because of the thermal output of the Sun. And in between, planets like Earth and Mars accumulated quartz (silicon dioxide), like the grains of sand on the beaches of the world, and a myriad of silicate compounds of granites and basalts like feldspars, pyroxenes, amphiboles, micas, clays, olivine and a myriad of others typical of the Earth's crust.

Hot radioactive short-lived elements, together with the intense heat from an increasing number of deadly collisions, turned Planet Earth into a cauldron of molten lava. Larger and larger planetesimals continued to crash destructively into the emergence of Planet Earth. And some think that the Moon was formed by one of these early collisions with *"Theia"*.[164] Shortly after its

---

164 Choi, Charles Q. 2019. The Moon May Have Formed When Earth's Magma Was Blasted into Space. Space.com.

formation, the lunar surface testifies that the Moon went through a period referred to as ***The Late Heavy Bombardment: A violent Assault on Young Earth.***[165] It is thought that the Moon, Earth and the inner planets of Mercury, Venus, and Mars all suffered from violent impacts during this early period. All but Earth wear the evidence of these impacts. Why? Earth soon erased it's scars by the recycling of its surface through plate tectonics and erosion.

Differentiation and the Great Iron Catastrophe

Like the refiner's fire, the Earth began its process of chemical differentiation. As temperatures within primordial Earth rose above the liquidus (melting point) for iron (1538 degrees centigrade), liquid globs of iron, along with other heavy metals like nickel (mined at the Sudbury impact site), migrated by gravitational forces to the center of the Earth. This event, the ***Great Iron Catastrophe***, was a major turning point in the molding of the planet because iron comprises a major fraction of Earth's mass (35%)[166] and is abundant in the Earth's core. Conversely, lighter slag-like compounds comprised predominately of silicon, oxygen, aluminum, sodium, and potassium migrated to the surface. The upward flow was encouraged by rapidly moving gases of sulfur, nitrogen, and hydrogen (water, methane) that were driven to the surface as the dense metals sank to the core. ***The core of the Earth remained as liquid metal until another major turning point took place; right before the beginning of the Cambrian.***

Primary Stratification of Planet Earth

Scientists know that the Earth has been undergoing differentiation; that is, becoming "chemically and mineralogically" stratified with time. This stratification on a planetary scale has resulted in the separation of the Earth into its crust (~1 % of Earth's volume), the mantle (84 %), and the core (15 %). The separation of solids and

---

165 Redd, Nola Taylor. 2017. The Late Heavy Bombardment: A violent Assault on Young Earth. Space.com.
166 Arnold, Kylene, 2018. What Four Elements Make Up Almost 90% of the Earth? Sciencing.com

liquids in the early Earth led to the differentiation of a rocky mantle of silicate rocks like peridotites, separated from the metallic iron rich core. But what about the core itself?

The Earth's stratification was, and continues to be, partially driven by the radioactive decay heat of Uranium-238, Uranium-235, Thorium-232, and Potassium-40, all of which have half-lives ranging from greater than 700 million years up to about 14 billion years. This radioactive decay heat helped keep the Earth's early core in a molten liquid state for most of the planet's existence. But even late in the Precambrian, scientists postulate that the primitive separation of the core into its inner and outer layers was sluggish or non-existent.

Miraculous Recovery of a Flickering Heartbeat of a Dying Patient

Long after the Great Iron Catastrophe which led to the formation of the Earth's primitive core, another key event took place within the core that would forever shape the Earth at the beginning of the Cambrian. It was a time when the heartbeat of the Earth's magnetic field was flickering and about to die forever. *But something miraculously took place to restore this most important heart*. The home to the lives of plants and animals and the home for mankind was being transformed about 6371 kilometers deep beneath our feet. At the beginning of the Cambrian period, *Earth's magnetic field nearly died during critical transition.* Scott Johnson reported that the work of a team at the University of Rochester,

> *found that the magnetic field was incredibly weak at the studied time {Early Cambrian} ... On top of that, it seems that the magnetic poles were flipping around extremely frequently. It's very weird behavior. Simulations have predicted that the solidification of the inner core was fairly recent in geologic terms, occurring right around {the Cambrian}. In those models, the reorganization of the core causes the magnetic field to go wild for a while, thrashing*

76

around in a weakened state. The researchers say their evidence fits with a scenario where the inner core only began solidifying around 565 million years ago—almost 4 billion years into the Earth's life. This adds to questions about how the magnetic-field-producing "geodynamo" in the core looked before this time and how it lasted so long. [167]

Events taking place in the deepest reaches of our planet's core were accompanied by rapid events on the surface of the Earth nearly 4000 miles above, as evidenced by the rapid movements of the Earth's magnetic poles, analogous to **atrial fibrillation.** Could these events involving the Earth's core and the Earth's crust be connected? Is it by coincidence that these events are connected with the sudden appearance of life and the re-emergence of a stronger magnetic field?

Diagnosis of weak fluctuating Magnetic Field → Heterogeneous Swirling Crystallite Fe clusters

According to Bono and Tarduno:[168]

These paleo-intensity values were 10 times less than the present magnetic field, lower than anything observed previously, Tarduno says. It suggested there's something fundamental going on in the core. Combined with previous studies that have found that the magnetic field was also rapidly reversing polarity during that time period, the new result indicates that Earth's field may have been on the point of collapse about 565 million years ago. That suggests that the inner core hadn't yet solidified. [169]

---

167 Johnson, Scott K. 2019. Earth's Magnetic Field Nearly Died During Critical Transition. arstechnica.com
168 Bono, Richard K. and John Tarduno. 2019. Young inner core inferred form Ediacaran ultra-low geomagnetic field intensity. Nature. Nature Geoscience 12, pages 143-147. www.nature.com
169 Gramling, Carolyn. 2019. Earth's Core may have hardened just in time to save its magnetic field. ScienceNews. www.sciencenews.org

Diagnosis: the core liquid was in the process of precipitating high density crystallites of iron-nickel. This early precipitation and settling of exceedingly dense heavy iron crystal aggregates were heterogeneously distributed within the deep embryonic inner core. The solid inner core rose from the precipitation of these crystals. As aggregated crystallites accumulated, they swirled and rotated in the liquid outer core; meanwhile, the Earth's crust was being violently bombarded. And the outer core imbalance resulting from these solid iron aggregates raised havoc with the Earth's magnetic field.

Heart Surgery in the Nick of Time

In the simplest sense, the initial molten center of the earth behaved like an iron foundry with **exceedingly dense heavy iron and nickel metal crystallites** forming under the intense core pressures and temperatures, settling to the center of the core while the liquid, less dense molten iron was forced to move outwards akin to the commonly known process of **magma filter pressing**, resulting in the separation of the primitive solid internal core from the outer liquid core. Filter pressing is defined as:

> *a process of magmatic differentiation wherein a magma, having crystallized to a mush of interlocking crystals in liquid, is compressed by Earth movements and the liquid moves toward regions of lower pressure, thus becoming separated from the crystals.[170]*

Today, researchers project the pressure and temperature at the boundary between the inner and outer core is about 6000 degrees Centigrade and 330 GPa at a depth of 5200 kilometers. By comparison the pressure and temperature at the center of the innermost core approaches 7000 degrees Centigrade[171] and 364 GPa at a depth of about 6400 kilometers.

---

170 mindat.org/glossary/filter_pressing
171 RPI http://ees2.geo.rpi.edu/geo1/lectures/lecture15/interior_08.html
Downloaded. August 6, 2019.

Continued Core Evolution: Nuclear Reactors?

While the materials of the core continued to organize into a central solid sphere, lighter elements were squeezed out. Elements like silicon, sulfur, and oxygen migrated through the liquid outer core forming a third lighter, outermost liquid core layer.

> *Even though the bulk of the core is iron, researchers also knew it contained a small amount of lighter elements such as oxygen and sulfur. As the inner core crystallized over time, scientists think this process forced out most of these light elements, which then migrated through the liquid outer core...The speeds at which seismic waves traveled through the outer core at different depths suggest that its composition does not remain the same all the way through. Instead, the uppermost 185 miles (300 km) or so is a distinct structure, with the section nearest the boundary consisting up to 5 percent by weight of light elements.*[172]   According to the authors, *this estimation favors a high heat flux at the core-mantle boundary with a possible partial melting of the mantle. <u>The temperature difference between the mantle and the core is the main driver of large-scale thermal movements, which together with the Earth's rotation, act like a dynamo generating the Earth's magnetic field</u>*[173] *<u>...centrifugal forces would have concentrated heavy elements like thorium and uranium on the equatorial plane and at the Earth core-mantle boundary. If the concentrations of these radioactive elements were high enough, this could have led to a nuclear chain reaction that became supercritical, causing a nuclear explosion.</u>*[174]

Are core-mantle nuclear explosions cyclical? Is it possible that nuclear materials buildup, explode, scatter, and re-accumulate?

---

172 Choi, Charles, Q. 2010. Earth's Core Has Another Layer, Scientists Claim. LiveScience. www.livescience.com and freerepublic.com/focus/
173 Zhang, Youjun and others. 2016. Experimental constraints on light elements in the Earth's outer core. Nature. *Scientific Reports* **6,** Article number: 22473
174 Edwards, Lin. 2010. The Moon may have formed in a nuclear explosion. phys.org.

9   Salt Deposits: Conundrum for Young Earth

The waters are hid as [with] a stone, and the face of the deep is frozen.[175]   I will give thee the treasures of darkness, and hidden riches of secret places, that thou mayest know that I, the LORD, which call [thee] by thy name, [am] the God of Israel.[176]

Salt layers, found throughout the geologic column pose a great conundrum for a literal Young Earth Creationist. The current theory of the origin of salt deposits is by seawater evaporation. But in many places, single salt deposits are hundreds of feet thick. If the current oceans were to completely evaporate, they would only leave behind a layer of salt several hundred feet thick. Yet the geologic column has numerous salt deposits from the Cambrian to the top, that in sum exceed thousands of feet. So if numerous thick layers are found at numerous intervals of time, just since the Cambrian, how many oceans would have to evaporate to form them? But the most important question is: how much time would it take to evaporate these multiple oceans? Evaporation may be viable for thin salt deposits but it is inadequate to explain the origin of numerous thicker (>1km) deposits.[177] The fact is, evaporation of seawater is a proven mechanism for the formation of salt beds. *We need only point to the Great Salt Lake within which its dissolved constituents are common in seawater including sodium and chloride ions, sulfates, and magnesium, calcium, and potassium.* *But a staggering amount of time would be required to form all of the world's salt deposits by evaporation.*

---

175 Job 38:30.
176 Isaiah 45:3.
177 Scribano, Vittorio and others. 2017. Origin of salt giants in abyssal serpentinite systems. Int'l Journal of Earth Science. 106(7): 2595-2608.

Seawater, Supercritical fluids, and Gases Recycle between the Upper Mantle and the Crust

Water, the essential ingredient for life, is hidden within the rocks found deep within Earth's interior by chemical bonds and as inclusions within the minerals found there. And within the Great Deep, we've learned that some of this water is present as a form of highly compressed "ice" referred to as Ice VII found in diamonds. The waters of the deep are essential in bringing ore minerals up to the Earth's crust by hydrothermal systems and processes. These heated hydrothermal waters combine with gases to form acids that dissolve rock constituents and transport them into the upper reaches of magma chambers and outward into faults and fractures in the Earth's crust. Through this thermal separation process, metals such as copper, lead, zinc, and gold are concentrated. They are segregated in the magma because these cations do not readily fill the crystal lattices of typical rock forming minerals. In the same manner, rare elements, like uranium and thorium are transported and brought to the Earth's surface for use by mankind.

The Earth's Upper Mantle and Crust are the products of differentiation and recycle, analogous to the differentiation of the Earth's Core. The uppermost layers of Earth can be thought of as three layers: the Transition Zone, the uppermost Mantle, and the Crust. These three layers far beneath us are connected by the recycling of matter important to life. The seafloor is pushed from the oceanic ridges and dives into the abyss by subduction at the oceanic trenches. As it descends, the ever increasing pressures squeeze water out of the descending rock. In turn, the liberated water lowers the temperature at which the sedimentary rock can melt while being carried downward. Likewise, upper mantle rock is melted, forming rising magmas that release their payloads of water, gases, enriched light elements, and exotic heavy elements attached to anions of carbon, hydrogen, sulfur, nitrogen and chlorine.[178] Within these systems, gases from these rising magmas mix with the descending seawater, forming acids and sulfates with the tendency to concentrate elements into ore bodies.[179]

Seismic Tomography proves that Ocean Floor Descends into the Lower Mantle by Subduction

The subduction of the ocean floor can be seen using seismic tomography. In essence, it's like using an ultrasound to see inside a human. As a prime example, the Farallon plate continues to be subducted under the North American Plate. Initially, the Farallon Plate pushed against the North American plate at a low angle at the time when the continent of Pangea was breaking up. During this compressional period, the Farallon plate participated in the thickening of the leading edge (west coast) of the North American Plate. This thickening is evidenced by the formation of the Rocky Mountains. However, the dynamics changed when the North American Plate was pushed up and over the Pacific Ocean Ridge (as seen on the back cover). The tectonics shifted, giving rise to the San Andreas transform fault and shearing along the continental margin. Some believe that from this point *the North American Plate began extending towards the west* with "stretch marks" evidenced by *the Basin and Range Province as the most obvious signature of expansion*.[180] Seismic tomography shows that the Farallon plate dove at about a 45 degree angle as it crossed through the upper mantle Transition Zone and continued its journey until it reached the core-mantle boundary, some 2700 kilometers below the surface where it arrived at the slab graveyard. [181]

---

178 Fischer, Tobias P. and Giovanni Chiodini. 2015. The Encyclopedia of Volcanoes (Second Edition). Chapter 45 – Volcanic, Magmatic, and Hydrothermal Gases. Academic Press.

179 Ibid. Cornel E.J.de Ronde and Valerie K.Stucker Chapter 47 – Seafloor Hydrothermal Venting at Volcanic Arcs and Backarcs. Academic Press.

180 Monroe, Robert. 2010. How the West was Stretched. Scripps News. UC San Diego, scripps.ucsd.edu

181 Stephen P. Grand, Rob D. van der Hilst, Sri Widiyantoro (1997): *Global Seismic Tomography: A Snapshot of Convection in the Earth.* GSA Today 7(4):1–7. The above image file is licensed under the Creative Commons Attribution-Share Alike 4.0 International license.

Water and CO_2 help melt Mantle Rock with substantial implications for Plate Tectonics

We have known for decades that when high concentrations of water, sodium, potassium, and/or carbon dioxide are added to silicate rocks, like those in the Earth's mantle, melting occurs at much lower temperatures.[182] That is, if you want to lower the melting point of a rock, add water, carbon dioxide, and pressure. The first phase diagram shows that by adding water to a mantle rock (in this case peridotite), the melting point drops by as much as 400^0C at 180 km. Likewise, as seen in the second diagram, when CO_2 is added to the rock, its melting point will drop more than 300^0C at a depth of about 150 km.

Depression of Melting Temperature of Mantle Rock by Water

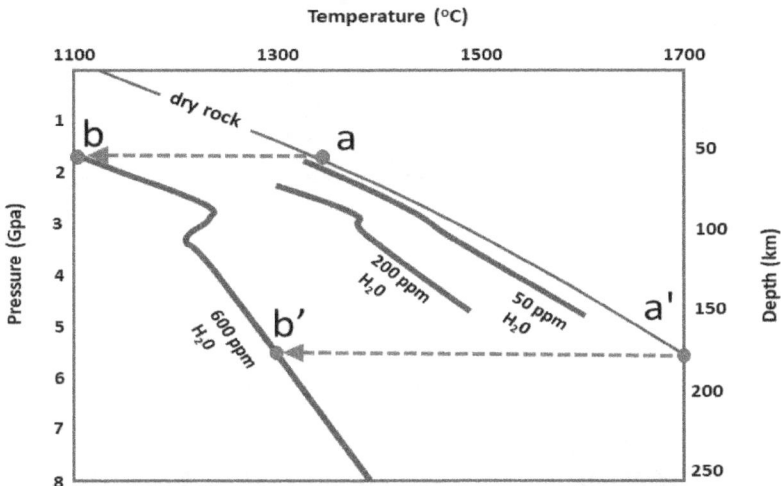

At a depth of 50 km 600ppm water drops the melting point of mantle rock by about 250^0C

At a depth of 180 km 600ppm water drops the melting point of mantle rock by about 400^0C

The buoyant rise of magma at the Mid-Ocean Ridges is key to the convection cycle that drives the plates under the continents. Importantly, water and CO_2 lower the melting temperature of rocks under oceanic ridges by several hundred degrees (300 to 400

---

182 NI HuaiWei, ZHANG Li and GUO Xuan. 2016. Water and partial melting of Earth's Mantle. SCIENCE CHINA Earth Sciences, 59 (4): 720-730.

degrees C) as compared to dry rock and therefore, are key to the recycling of Earth's crust. As a point of interest, rocks known as carbonatites, like those of McClure Mountain, contain high amounts of water, carbonate, sodium, and potassium ions. When evaluated in the laboratory, researchers find that these rare rocks melt at much lower temperatures and pressures than typical seafloor basalts. This demonstrates the importance of water and gases in the melting and fluidization of mantle rock.

Depression of Melting Temperature of Mantle Rock by CO_2

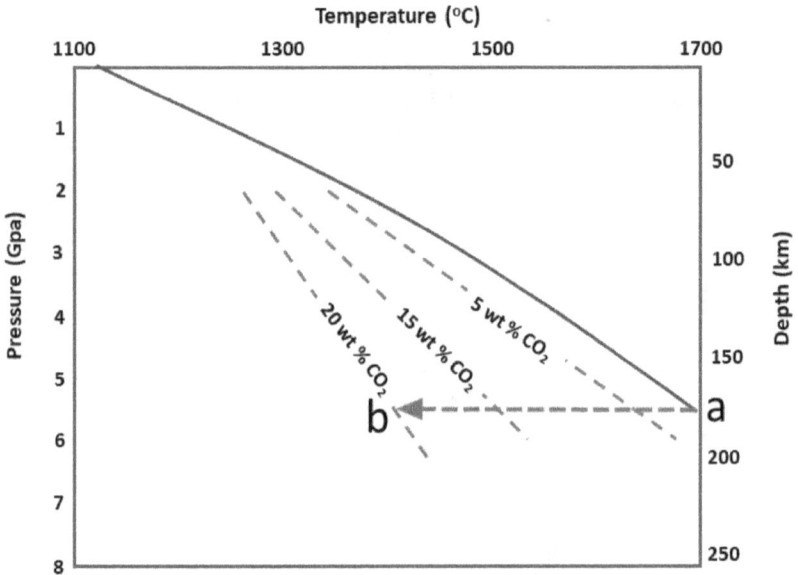

At a depth of 150 km 20 wt.% CO2 drops the melting point of mantle rock by about 300°C

Correlation of Tectonic Zones, Serpentinization, and Salt Deposits

Cambrian salt deposits are found in the Pakistan Salt Range, in salt basins of the Middle East including the Jordan Valley and the Dead Sea, in Eastern Russia, and in China to name but a few localities. *These deposits, like numerous others since the Cambrian are frequently associated with tectonic zones.* Numerous authors are beginning to realize the importance of this correlation. Significantly, the correlation to tectonic zones extends beyond salt deposits and includes correlations to hydrocarbon resources[183] and mineral deposits.[184] Many of these types of economic deposits are associated with the mantle-crust recycle,[185] discussed above, where enrichment takes place due to the action of water and gases, especially carbon dioxide. Numerous authors even suggest that upper mantle serpentinized peridotite is a common brine source. According to Stanley Keith and others:[186]

Dehydration of the serpentinite source to talc (steatization) by mantle heat during failed, intra-continental rifting of the Pangea supercontinent at the end of Permian time released vast amounts of element-laden, high density brines into deep-basement fractures...

In this case, serpentinization is a process in which common mantle rock is altered by seawater. When the peridotite is moved nearer to the surface by tectonic forces, the presence of fluids and the reduction in pressure and temperature (about 400^0C) transform minerals like olivine into serpentine. According to a report by NOAA,[187] the chemical reactions between seawater and these upper mantle rocks has several important consequences:

the consequence of the formation of serpentine during hydration of mantle rocks is that the density of the rocks change from about 3.3 grams per cubic centimeter (g/cm3) to about 2.7 g/cm3 and the volume of the rocks can increase

183 Ghori, K.A.R., Jonathan Craig, Bindra Thusu, and Sebastian Luening. 2009. Global Infracambrian petroleum systems: a review. Geological Society of London, Special Publications, volume 326. Pages 109-136.

184 Zheng, Yongfei and others. 2019. Hydrothermal ore deposits in collisional orogens. Science Bulletin 64, 205. Science China Press.

185 Sun, Weidong and others. 2015. Subduction and Ore Deposits. International Geology Review 57 (9-10):iii-vi.

186 Keith, Stanley B., Volker Spieth, and Jan C. Rasmussen. 2017. Zechstein-Kupferschiefer Mineralization Reconsidered as a Product of Ultra-Deep Hydrothermal, Mud-Brine Volcanism. Chapter 2. DOI: 10.5772/Intechopen.72560

187 Früh-Green, Gretchen. (Revised August 2010.) The Lost City 2005 Expedition. Oceanexplorer.noaa.gov.

by 20-40%. Serpentinization also affects other geophysical properties of the oceanic crust by lowering the seismic velocities of the rocks, changing their gravity signatures and mechanical properties, and by <u>increasing their degree of magnetism</u>. The change in density and the expansion of the rocks during serpentinization has the important effect that the mountain <u>becomes lighter and needs more room as it swells up, and thus "lifts" itself to greater elevations</u>... This in turn creates new fractures that allow seawater to penetrate further...

The volume expansion due to serpentinization is a key to the fracturing of the crust! The reaction of seawater with the minerals, such as olivine, in mantle rocks consumes significant volumes of water and releases salt.

Serpentinization of Mantle Rock leads to Massive Salt Formation and Storage of Water and CO_2

The chemical reaction processes of converting peridotite (mantle rock) to serpentine is accompanied by a whole host of important chemical reactions in steps 1-3 listed below.[188] Numerous other chemical reactions between other gases in seawater and rock are important to our emerging story:

1. Mafic minerals in the mantle rock consume water and produce a *concentrated salt solution*.
2. Seawater removes carbon dioxide from the atmosphere and from submarine volcanic gases. The reaction between carbon dioxide enriched seawater and mafic minerals in mantle rock also *stores carbon dioxide* (process of sequestration).

---

188 Ueda, Hisahiro and others. 2016. Reactions between olivine and CO2-rich seawater at 300 C:Implications for H2generation and CO2sequestration on the early Earth. Geoscience Frontiers. Elsevier.com

3. The reaction of carbon dioxide and mantle rock produces *limestone and dolomite* as a part of the process.
4. Emissions of sulfur gases by submarine volcanic emissions, demonstrated under experimental conditions with alkali and alkali earth metals (sodium, calcium, and magnesium), form respective *sulfates*.[189]
5. Admixtures of carbon dioxide and sulfur gases result in the precipitation of both carbonate and *sulfur-bearing minerals including anhydrite, alunite, and pyrite*.[190]
6. Could Banded Iron Formations also be a by-product of peridotite (feldspar and pyroxene) reactions with seawater which produce silicon dioxide[191] and iron oxide?[192]
7. Serpentinization of peridotites yields hydrogen upon oxidation of ferrous iron to ferric iron by seawater reaction.[193]

...the deposition of huge amounts of marine salts, including the formation of tens of metres of highly soluble types (tachyhydrite and bischofite) ...during the Messinian Salinity Crisis, are inconsistent with the wet and warm palaeoclimate conditions .... Our results indicate that salt and brine formation occurs during serpentinization and that the brine composition and salt assemblages are dependent on the temperature and CO_2 partial pressure. Our findings help explain the presence and sustainability of highly soluble salts that appear inconsistent with reconstructed climatic conditions and demonstrate that the presence of highly

189 Chareev, Dimitriy A. and others. 2009. Experimental Study of Sulfur Dioxide Interaction with Silicates and Aluminosilicates at Temperatures of 650 and 850°C. Geochemistry International, 2010, Vol. 48, No. 10, pp. 1039–1046.
190 J. L. Palandri and Y. K. Kharaka, "Ferric Iron Bearing Sediments as a Mineral Trap for CO2 Sequestration: Iron Reduction Using Sulfur Bearing Waste Gas," Chem. Geol. 217, 351–364 (2005).
191 Chareev, Dimitriy A. and others. 2009.IBID.
192 Huang, Ruifang and others. 2017. The production of iron oxide during peridotite serpentinization: Influence of pyroxene. Geoscience Frontiers Volume 8, Issue 6, November 2017, Pages 1311-1321.
193 Klein, Frieder and Wolfgang Bach. 2009. Fe–Ni–Co–O–S Phase Relations in Peridotite–Seawater Interactions. *Journal of Petrology*, Volume 50, Issue 1, Pages 37–59.

soluble salts probably has implications for global tectonics and paleoclimate reconstructions.[194]

Origin of Thick Salt Deposit by Upper Mantle Precipitation from Supercritical Fluids

Rift basins throughout the Middle East are sites of Cambrian salt deposits in subsiding rift environments along the Middle Eastern edge of Gondwanaland.[195]

Extremely low solubility of typical seawater salts within certain supercritical sections of their pressure-temperature composition space is a proven experimental fact. Its consequences are often referred to as either 'shock crystallization' or 'out-salting'. <u>Our alternative model for the formation of salt deposits hypothesizes that high temperatures and pressures characteristic for the high heat-flow zones of tectonically active basins may bring submarine brines into the out-salting regions and result in the accumulation of geological- scale salt depositions.</u>[196]

The circulation of salt-rich seawater in rapidly subducting slabs and along rift valleys provides an environment for *massive salt deposits, likely derived from the action of both supercritical fluids and serpentinization. Both are dependent on temperature and CO_2 concentrations.* Could it be, that far more massive salt beds worldwide, are the product of supercritical fluid precipitation? You be the judge. *I believe these mechanisms resolves a major problem for the age of the Earth we refer to as the Phanerozoic (Cambrian to Present).*

---

194 Debure, M. and others. 2019. Thermodynamic evidence of giant salt deposit formation by serpentinization: an alternative mechanism to solar evaporation. Scientific Reports, Volume 9, Article Number 11720.
195 Moujahed I. Husseini and Sadad I. Husseini. 1990. *Origin of the Infracambrian Salt Basins of the Middle East.* Geological Society, London, Special Publications, 50, 279-292.
196 Hovland, M. and others. 2006. Basin Research. Volume 18. Pages 221-230.

10 *Water, CO₂, and Salt: Greasing the Skids*

In his hand [are] the deep places of the earth:[197] the earth, shall shake..., and the mountains shall be thrown down, and the steep places shall fall, and every wall shall fall to the ground.[198] The waters thereof roar [and] be troubled, [though] the mountains shake with the swelling thereof.[199] And the mountains shall be molten under him, and the valleys shall be cleft, as wax before the fire[200] ... The mountains quake at him, and the hills melt, and the earth is burned at his presence[201]

Can the Earth's plates be moved rapidly, and perhaps catastrophically? Does Nature provide a means of lubricating the plates so that frictional forces can be overcome? What forces would it take to move mountains? The Himalayas have ten of the Earth's fourteen peaks that exceed 8000 meters (26,000 feet) like Mount Everest of Nepal and China. Even at the top of Mount Everest, which has a peak of 29,035 feet, rock is found with fossils of sea creatures that once lived at the bottom of the ancient Tethys Sea. Everest rose because of an enormous collision between the Indian and Eurasian plates, carrying sea lilies (crinoids) from the Cambrian seafloor. And yes, these fossils are found living today in the oceans at depths of 200 meters. But what about frictional forces that had to be overcome? Were the Himalayas raised rapidly or at centimeters per year as advertised? Scientists are finding that lubrication can overcome frictional forces and enable rapid plate movement.

---

197 Psalm 95:4.
198 Ezekiel 38:20.
199 Psalm 46:3.
200 Micah 1:4.
201 Nahum 1:5.

Heart Mountain Klippe: High CO_2 Drastically Reduced Friction, causing a runaway Mountain

Heart Mountain, just north of Cody, Wyoming is what scientists refer to as a tectonic klippe. The block slide covered an area of over 3400 square kilometers and moved between 27 miles and up to 62 miles. Like a runaway freight train, the 8000 foot mountain (prominence of 2163 feet) raced northwest across the Wyoming landscape. At the end of its journey, it came to rest upon rocks 300 million years younger than the rocks that it carried. But what propelled this mountain into motion? The Laramide Orogeny caused the uplift of the Beartooth Mountain Range about 50 to 75 million years ago and at the same time the Earth's crust was being buckled, crumpled, and forced downward forming the Bighorn and Absaroka Basins. The formations of these mountains and basins was followed and accompanied by volcanic eruptions. Several hypotheses have been advanced to explain the rapid sliding of the mountain over younger ground. All proposed mechanisms include one form of lubricant or another. The younger rock surface is comprised of **limestone which would break down by friction, releasing <u>carbon dioxide gas</u>** which would **"float" the sliding mountain like an air pallet.**

We suggest that frictional heating led to dissociation of carbonate rock along the fault, producing supercritical CO_2 as the suspending medium. <u>High CO_2 pressure drastically reduced friction</u> along the fault and allowed continuation of catastrophic movement, probably initiated by a volcanic or phreatomagmatic explosion, resulting in very large displacement on a low-dipping surface. Earlier slower sliding may have occurred but final emplacement was rapid (minutes) and spectacular.[202]

---

202 Beutner, Edward C. and Gregory P. Gerbi. 2005. Catastrophic emplacement of the Heart Mountain block slide, Wyoming and Montana, USA. Geological Society of America Bulletin, v. 117, 724-735.

Pseudotachylyte is found at the base of the mountain which is commonly found in impact sites like Vredefort, South Africa and Sudbury, Ontario and at the base of large landslides. Pseudotachylyte[203] is formed from the breakdown of glass, comprised of reactions along the slide surface. Based on their studies, Ahronov and Anders conclude that the Heart Mountain block travelled as much as 62 miles in 30 minutes![204] *Fluid over-pressuring and supercritical CO_2 coupled with volcanic activity were all likely participants in moving Heart mountain.* This glassy matrix is the <u>melt product of the heat of friction</u>.

> *Dike injections increased horizontal stresses and heated the surrounding layers. Both the increased stresses and the heat input elevated fluid pressure of water trapped within the [Big Horn Dolomite] BHD. In addition, vertical hydrofracturing was retarded as horizontal stress approached vertical, thus allowing a critical buildup of fluid pressure. Fluid over-pressuring is a mechanism that can overcome the mechanical problem of initiating movement on a low-angle surface. Moreover, this mechanism explains the observed fluidized features found along the basal contact of the slide block as well as the observed lack of deformation in the lower plate.* [205]

Chief Mountain, Montana, ~170 MA BP

Chief Mountain, another well studied klippe, is found in the adjacent State of Montana on the Eastern border of Glacier National Park. The mountain reaches an elevation of over 9000 feet and rises over the surrounding plains by 1840 feet. Chief

---

203 Legros, F., Cantagrel, J-M. & Devouard, B. 2000. Pseudotachylyte (Frictionite) at the Base of the Arequipa Volcanic Landslide Deposit (Peru): Implications for Emplacement Mechanisms. The Journal of Geology, volume 108, p. 601–611.
204 Binns, Corey. 2006. Landspeed Record: Mountain Moves 62 Miles in 30 Minutes. Livescience.com.
205 Aharonov, Einat and Mark H. Anders. 2006. Hot water: A solution to the Heart Mountain detachment problem? Geology. 34 (3): 165-168.

Mountain is comprised of ~600 million year old Precambrian rock resting upon ~100 to 200 million year old Cretaceous rocks. As a result of the Sevier/Laramide Orogeny which coincided with the breakup of Pangea, a slab of Belt Series rocks, hundreds of miles wide and 15-18 miles thick, was thrust over the Cretaceous shale bearing formations more than 50 miles east! The block slid upon a décollement surface[206] which is a gliding plane between two rock masses, also known as a basal detachment fault. The basal fault is widely known as the Lewis Overthrust which extends north into Alberta, Canada.

Salt Range Décollement: Moving Mountains

A massive layer of halite (salt) forms the base of the Salt Range Mountains of Pakistan. The salt acts as the surface upon which the massive mountain chain is transported by tectonic forces acting between the Indian and Eurasian plates. Salt is a well-known "lubricant" that reduces friction along basal thrust faults. This type of surface is a gliding plane between two rock masses, also *known as a basal detachment fault or décollement.*

---

206 Simony, Philip S.; Carr, Sharon D. (2011-09-01). "Cretaceous to Eocene evolution of the southeastern Canadian Cordillera: Continuity of Rocky Mountain thrust systems with zones of "in-sequence" mid-crustal flow". *Journal of Structural Geology*. **33** (9): 1417–1434

Water, Pseudotachylyte, CO_2 and Salt Grease Plate Tectonics

Experiments demonstrate that *water, carbon dioxide, and salt decrease the viscosity of upper mantle rocks by two orders of magnitude or more thereby increasing heat convection and creep rate (allowing them to flow more readily).* These constituents, greatly reduce friction, enabling plate tectonics:[207]

> *The transition to plate tectonics is controlled by the viscosity ratio between the lithosphere and a convecting mantle. Thus, the steady thickening and stiffening of the lithosphere through cooling early in the planet's evolution must be halted before the lithosphere becomes too stiff, and at the same time, mantle viscosity has to be low enough to encourage convection.[208] Experimental results indicate that the viscosity of olivine aggregates is reduced by a factor of ~ 140 in the presence of water at a confining pressure of 300 MPa and that the influence of water on viscosity depends on the concentration of water in olivine.[209]*

Serpentinization of mantle rock dramatically increases its ability to flow. And the same processes lead to lower melting temperature, thereby increasing admixtures of rock + melt buoyancy and ability to flow! And the density of serpentinized peridotite is considerably less dense than the unaltered rock. At the same time, the alteration leads to a significant increase in mantle rock volume. Reaction of mantle rock with CO_2 is even more dramatic:

---

207 Kono, Yoshio and others. 2014. Ultralow viscosity of carbonate melts at high pressures. Nature Communications 5, Article number: 5091.
208 Tikoo, Sonia M,. and Linda T. Elkins-Tanton. 2017. The fate of water within Earth and super-Earth's and implications for plate tectonics. Philosophical Transactions A Math Phys Eng Sci 28; 375 (2094).
209 Hirth G, Kohlstedt DL. 1996. Water in the oceanic upper mantle: implications for rheology, melt extraction and the evolution of the lithosphere. Earth Planet Sci. Lett. 144, 93–108.

Carbonate melts possess lower densities than basalt melts, further enhancing the contrast in melt mobility between basalt and carbonate melts beyond the significant differences in viscosity. As a result, the mobility of carbonate melts is $101–148\,g\,cm^{-3}\,Pa^{-1}\,s^{-1}$ at 30–180 km depths, which is ~<u>200–1,700 times higher than those of the basalt</u> ... These data suggest that, while basalts are erupted at mid-ocean ridges, the <u>initial melts generated at depth in this environment may be carbonate rich</u>, and that they progressively change to basaltic composition during ascent and melt-rock reaction.[210]

Kimberlites, carbonatites, and lamproites, like those that I sampled and analyzed in the McClure Mountain Complex of Colorado were demonstrated to be the result of differentiation and separation of the carbon dioxide rich magmas from basaltic magmas based upon extensive chemical analyses and modeling that I conducted.[211] The separation of the fluids as pseudo-immiscible fluids provides a water and gas rich forward-penetrating fluid that opens fractures which facilitates the rapid ascent of these magmas into the Earth's crust. These characteristics are likely key to initiation and acceleration of plate movement. *Water and CO_2 substantially reduce the melting point of mid-ocean ridge basalt and increase its creep rate. Together, their presence allows a rapid rate of plate tectonics. At the same time, salt concentrated by the serpentinization of mantle rocks serves as a lubricant that facilitates subduction and obduction (thrusting of large tectonic plates over others).*

---

210 Kono, Yoshio and others. 2014. Ultralow viscosity of carbonate melts at high pressures. Nature Communications volume5, Article number: 5091
211 Alexander, D.H. 1981. Geology, Mineralogy, and Geochemistry of the McClure Mountain Alkalic Complex, Fremont County Colorado. PhD Dissertation. The University of Michigan. 327 pages.

11 *Cambrian Explosion: In the Twinkling of an Eye*

And the earth was without form, and void[212]… And God said, Let the waters under the heaven be gathered together unto one place, and let the dry [land] appear: and it was so.[213] And the earth brought forth grass, [and] herb yielding seed after his kind, and the tree yielding fruit, whose seed [was] in itself, after his kind[214]… And God said, Let the waters bring forth abundantly the moving creature that hath life[215]…The hearing ear, and the seeing eye, the LORD hath made even both of them[216] in the twinkling of an eye.[217] From the beginning were eyewitnesses… of the word[218] [and] eyewitnesses of his majesty.[219]

E arth was readied for life, just as a terrarium is prepared for its inhabitants. ***There was the Day*** when the Earth appeared without form and void. It was a time when the Earthly terrarium was shrouded in a strange darkness, as the gases and ashes of the early atmosphere separated light from darkness. ***There was the Day*** when the waters on the surface were separated from the vapors that arose into the heavens to form a canopy. And ***There was the Day*** when the dryland appeared. ***There was the Day*** when plants would inhabit the land. Eons of time had passed since the Late Heavy Bombardment when planet Earth was

---

212 Genesis 1:2.
213 Genesis 1: 9, 10.
214 Genesis 1:12.
215 Genesis 1:20.
216 Proverbs 20:12.
217 1 Corinthians 15:52.
218 Luke 1:2.
219 2 Peter 1:16.

formless and void. The dross of the molten planet rose to the surface as silicate melts of granites, granodiorites, and their cousins floundered upon the magma sea like giant rafts. These comparatively low density bergs of silicate rock floated upon a once molten sea of higher density magnesium and iron rich magma which quenched to form the ocean floors. Since the early Precambrian, a primitive ocean was forming which reacted with the molten magma, to produce an ever thickening crust of quenched rock much like the basalt that rolls from Kilauea, Maunaloa, Hualalai and Maunakea today.

First Land Plants Pave the Way for the Cambrian Explosion

Many authors thought that there was no life on Rodinia and its fragmented landmasses. The oldest known plant fossils were thought to be on the order of 480 million years old. But genetic studies by one group of researchers suggest that lichen-like land plants appeared about 700 million years ago while fungi appeared about 1300 million years ago,[220] well before the *Cambrian Explosion of animal life*.

The early appearance on the land of fungi and plants suggests their plausible role in both the mysterious lowering of the Earth's surface temperature during the series of Snowball Earth events roughly 750 million to 580 million years ago and the sudden appearance of many new species of fossil animals during the Cambrian Explosion era roughly 530 million years ago. "Both the lowering of the Earth's surface temperature and the evolution of many new types of animals could result from a decrease in atmospheric carbon dioxide and a rise in oxygen caused by the presence on land of lichen fungi and plants at this time, which our research suggests," Hedges says. "An increase in land plant abundance may have occurred at the time just before the

220 Heckman, Daniel S.,David M, Geiser and others. 2001. Molecular Evidence for the Early Colonization of Land by Fungi and Plants. Science. Vol. 293, Issue 5523, pp. 1129-1133.

period known as the Cambrian Explosion, when the next Snowball Earth period failed to occur because temperatures did not get quite cold enough...I suspect the biggest cooling effect came from the reduction of carbon dioxide in the atmosphere by fungi and plants, which we have shown were living on the land at that time." Hedges says. "The plants conceivably boosted oxygen levels in the atmosphere high enough for animals to develop skeletons, grow larger, and diversify."[221]

Based on the study, the authors conclude that *molecular clock estimates* suggest much earlier colonization of the planet by plants. Their protein sequence analyses indicate that green algae and major lineages of fungi were present a billion years ago and that land plants appeared about 700 million years ago. Plants would have provided food and hiding places for the first animals. The teams later work suggests that *mitochondria and organisms with more than 2–3 cell types appeared soon after the initial increase in oxygen levels at 2300 Ma. The addition of plastids at 1500 Ma, allowing eukaryotes to produce oxygen, preceded the major rise in complexity.*[222] Plants had established themselves in the late Precambrian,[223] and through the process of photosynthesis they consumed massive quantities of carbon dioxide emanating from worldwide volcanism. In doing so, they produced large quantities of oxygen thus tipping the balance for animal life through the *Great Oxygen Event*. And perhaps as importantly, the early photosynthesis was instrumental in the precipitation of massive amounts of carbonates[224] of calcium and magnesium forming

---

221 Hedges, Blair and Barbara Kennedy. 2001. First Land Plants and Fungi Changed Earth's Climate, Paving the Way for Explosive Evolution of Land Animals, New Gene Study Suggests. Eberly College of Science, Penn State University. science.psu.edu.

222 Hedges, S. Blair and others. 2004. Ibid.

223 Heckman, Daniel S., Geiser, David M. and others. 2001. Molecular Evidence for the Early Colonization of Land by Fungi and Plants. Science. Vol. 293, Issue 5523, pp. 1129-1133.

224 Kosamu, I.B.M. and M. Obst. 2009. The influence of picocyanobacterial photosynthesis on calcite precipitation. International Journal of Environmental Science & Technology, Volume 6, Issue 4, pp 557–562.

limestones and dolomites. At the same time, iron was precipitated by oxygen produced by these plants and bacteria, generating massive banded iron formations.[225] Concurrently, plate tectonics led to the breakup of Rodinia rotating the newly forming shores of Gondwana, Laurentia, Baltica, Siberia, and the offshore land mass of proto-China to warmer latitudes which also encouraged the precipitation of the carbonate sediment beds.

Conditions Favorable for the Onset of Animal Life on Continental Shelves

And so it was that ocean bottom dwelling animals took their residence along the coasts of the shallow warm water seas, in the time referred to as the ***Cambrian Explosion.*** Over the last several decades, the scientific literature is mixed on whether oxygen was the key factor in triggering life.[226, 227] Several theories have been advanced to account for sufficient dissolved oxygen to support life: 1) ***first hypothesis-*** only in the Cambrian did the numbers of oxygen-depleting bacteria reduce in numbers sufficiently to permit the high levels of dissolved oxygen we know today;[228] 2) ***second hypothesis-*** a large impact in the ocean could have blown a large portion of the ocean into the atmosphere resulting in oxygenation;[229] and 3) ***third hypothesis-*** fungi appeared on land about 1300 million years ago.[230] As noted by several authors, conditions for life include far more than adequate dissolved oxygen. Numerous other environmental conditions are essential

225 Fru, Ernest Chi, and others. 2013. Fossilized iron bacteria reveal a pathway to the biological origin of banded iron formation. Nature Communications.
226 Joel, Lucas. 2017. When Oxygen disappeared, early marine animals really started evolving. Science Magazine. sciencemag.org.
227 Matthew R. Saltzman, Seth A. Young, Lee R. Kump, Benjamin C. Gill, Timothy W. Lyons, and Bruce Runnegar. 2011. Pulse of atmospheric oxygen during the late Cambrian. PNAS volume 108 (10) 3876-3881.
228 https://ucmp.berkeley.edu/cambrian/cambtect.html
229 Bontemps, Johnny. 2013. Did a Huge Impact lead to the Cambrian Explosion? Astrobiology Magazine. astrobio.net.
230 Heckman, Daniel S.,David M, Geiser and others. 2001. Molecular Evidence for the Early Colonization of Land by Fungi and Plants. Science. Vol. 293, Issue 5523, pp. 1129-1133.

for life on ocean shelves. These include sunlight, dissolved oxygen, salinity, pH, carbon dioxide, depth levels, and tidal action among others.[231]

Plate tectonics caused the breakup of Rodinia and the rotation of the plates to form Gondwanaland, Laurentia, Baltica, Siberia, and small fragments including China nearer the equator. And the rotation of the plates was accompanied by massive ocean ridge volcanism which released enormous volumes of carbon dioxide, water, and other gases to the oceans. Carbon dioxide is less soluble in warm water solutions than in cold water. Therefore, warming of shallow waters results in the precipitation of calcite and other carbonates. Because of the shift in the Cambrian landmasses to the equator, large carbonate platforms of limestone and dolomite were formed on warm continental shelves.

> *... creatures whose remains we find in the Burgess Shale lived in deeper (basinal) waters, making their homes on or in the sea floor or swimming above it. The local environment would have been calm - safely below the churning surface caused by storms or hurricanes... Most animals lived at the base of a large submarine cliff known as the Cathedral Escarpment. This formed at the outer edge of a wide, <u>tropical platform of carbonate rock</u> that may have extended as far as 400 kilometres (320 miles) from the shoreline. At least twelve fossil localities have been discovered at the foot of the Escarpment along a 60-kilometre belt running roughly north-south. This suggests the Escarpment might have helped optimize conditions for a rich animal community to develop and be preserved as fossils. The Escarpment itself was about 200 metres (650 feet) high before mud and other sediments began to fill in the basin. The shape*

---

231 Catherine E. Brennan, Hannah Blanchard, and Katja Fennel. 2016. Putting Temperature and Oxygen Thresholds of Marine Animals in Context of Environmental Change: A Regional Perspective for the Scotian Shelf and Gulf of St. Lawrence. PLoS One. 2016; 11(12): e0167411. Online 2016 Dec 20.

of the Escarpment may have channelled mudflows at its base, resulting in periodic deposits that enveloped and preserved the organisms living there.[232]

Similar extensive platforms of carbonate rock (limestones and dolomites) rimmed Gondwanaland and its associated fragmented land masses[233, 234] including Laurentia with its Burgess shale fossil beds, and the offshore land mass that provided homes for amazing life forms; later recognized as the China Chengjiang fossil beds.

Sauk Transgression: Erosional Formation of the Great Unconformity provided another Trigger for the Cambrian Explosion

Huge walls of water surged across Gondwanaland and the remaining remnants of Rodinia, tearing up the landscape, in a way similar to the Missoula Floods that eroded the channeled scablands of Eastern Washington State. Oceans of water relentlessly raced back and forth across the fragments of Rodinia grinding the lands to their crystalline cores. Worldwide action of peneplanation[235] was like a large grinding wheel as it cut the continents down forming a more or less broad horizon that geologists call the *Great Unconformity*. Some see the development of this Great Unconformity as yet another trigger of the Cambrian Explosion.

New stratigraphic and geochemical data show that early Palaeozoic marine sediments deposited approximately 540– 480 Myr ago record both <u>an expansion in the area of shallow</u>

232 Royal Ontario Museum. Downloaded 2019. The Burgess Shale. burgess-shale.rom.on.ca/

233 Elicki, Olaf. On-line since 2006. Microbiofacies analysis of Cambrian offshore carbonates from Sardinia (Italy): environment reconstruction and development of a drowning carbonate platform. Notebooks on Geology: Article 2006/01 (CG2006_A01).

234 James, Noel P., and David I. Gravestock. 1990. Lower Cambrian shelf and shelf margin buildups, Flinders Ranges, South Australia. Sedimentology. Volume 37, Issue 3.

235 A peneplain is a plain without hills or mountains formed by a long period of extensive erosion and often times considered in connection with tectonics.

epicontinental seas and anomalous patterns of chemical sedimentation that are indicative of increased oceanic alkalinity and enhanced chemical weathering of continental crust. These geochemical conditions were caused by a protracted period of widespread continental denudation [erosion] during the Neoproterozoic… The resultant globally occurring stratigraphic surface, which in most regions separates continental crystalline basement rock from much younger Cambrian shallow marine sedimentary deposits, is known as the Great Unconformity. Although Darwin and others have interpreted this widespread hiatus in sedimentation on the continents as a failure of the geologic record, this palaeogeomorphic surface represents a unique physical environmental boundary condition that affected seawater chemistry during a time of profound expansion of shallow marine habitats. Thus, the formation of the Great Unconformity may have been an environmental trigger for the evolution of biomineralization and the 'Cambrian explosion' of ecologic and taxonomic diversity following the Neoproterozoic emergence of animals. The Great Unconformity is well exposed in the Grand Canyon, but this geomorphic surface, which records the erosion and weathering of continental crust followed by sediment accumulation, can be traced across Laurentia and globally, including Gondwana, Baltica, Avalonia and Siberia, making it the most widely recognized and distinctive stratigraphic surface in the rock record. [236]

According to Shannon and Gaines, the Cambrian period saw the peak in the $^{87}Sr/^{86}Sr$ ratio, a minimum in neodymium, and a substantial increase in seawater concentrations of calcium $[Ca^{2+}]$. *They attribute these three chemical markers to the tectonism driven erosion and enhanced continental weathering during the emerging worldwide Great Unconformity.*

---

236 Peters, Shannan E., and Rebert R. Gaines. 2012. Formation of the 'Great Unconformity' as a trigger for the Cambrian explosion. Nature. Volume 484. Pages 363-366. Additional Supplementary Information 17 pages.

Oceans Flooded the Land during the Cambrian

Sea levels were nearly a thousand feet higher during the Cambrian and Ordovician than they are today. At the same time Cambrian landmasses were low lying, given that plate tectonics had not yet created the great mountain ranges seen today. Consequently, the oceans flooded the land with shallow water environments, and inland seas, providing a desirable environment for life.

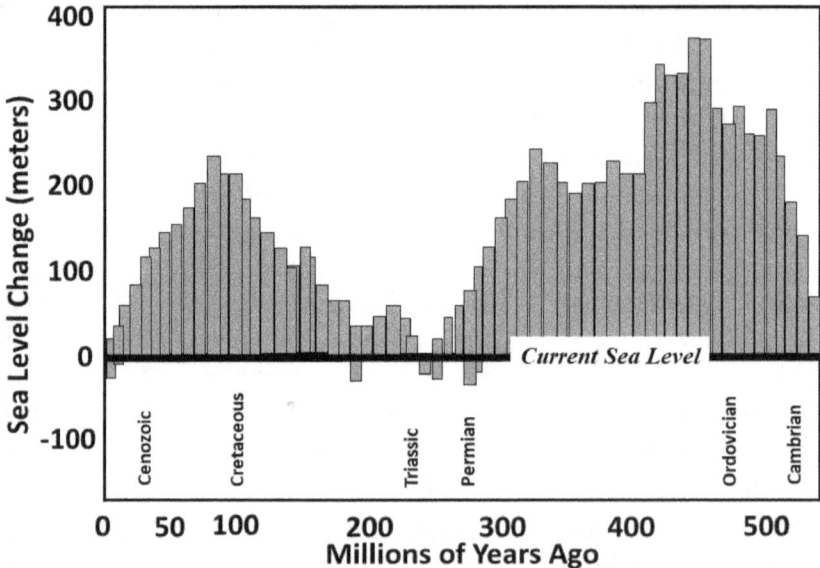

Absolute sea level changes (in meters) over the last 540 million years. Adapted from curves from Hallam and Exxon.[237]

Cambrian Life Phylogeny: Failure of the Evolutionary Concept

The fossil record of life forms preserved in the Cambrian Chengjiang and Burgess shale sites include many of the animal phyla that still exist to the present day. New Cambrian fossil

---

237 Hallam, A., *Phil. Trans. Royal Soc. B* 325, 437-455 (1989) and Exxon: Haq et al. 1987, Ross & Ross 1987, Ross & Ross 1988.

sites[238] are continuing to be discovered with a high degree of complexity.[239] *The notion of evolution advanced by Darwin is expressed in simplest terms by his "Tree of Life." However, numerous scientists see weaknesses in Darwin's proposition. Perhaps the single most unfavorable fact is the abruptness with which not one, not two, not three but tens of phyla appear in the Cambrian Period of the geologic record without the obvious links to predecessors. <u>Without predecessor links, the continuum of change that Darwin proposed is flawed and the leap to not one or two but virtually all sophisticated Cambrian phyla remains without a foundation.</u>*

Eyes, Brains, Digestive and Nervous Systems

Eyes are useless without a brain to interpret information transmitted by light rays. Eyes with a brain, but without a nervous system to communicate messages, are equally as useless. Anomalocaris had the eyes of a shrimp. By comparison, trilobites had large lateral compound eyes like the horseshoe crab. How is it that these two very different sea creatures suddenly appear from the abyss of time with different types of eyes? And what about the numerous other eye types that suddenly appeared at the same time period, such as the Opabinia regalis with its 5 compound eyes on stems found in the Cambrian Burgess Shale of British Columbia; and fully functioning eyes at that. Trilobite eyes are analogous to the compound eyes of insects (Arthropoda insecta) found in the world today. How is it that these fully functional "ancient" sea creatures appeared abruptly, without predecessors, in the seas of the Cambrian. How is it that their very eyes have not evolved significantly over hundreds of millions of years? *Their eyes were supported with brains and a nervous system, and fully functioning digestive systems, even with digestive glands,[240] and*

---

238 Weisberger, Mindy. 2019. Bonanza of Bizarre Cambrian Fossils Reveals Some of the Earliest Animals on Earth. livescience.com.

239 Gramling, Carolyn. 2019. Newfound fossils in China highlight a dizzying diversity of Cambrian life. ScienceNews. sciencenews.org.

240 Hopkins, M.J., F. Chen, S. Hu, and Z. Zhang. 2017. The oldest known digestive system consisting of both paired digestive glands and a crop from

reproductive systems. How is it that these body systems, found in insects today, were fully present in these creatures from more than 500 million years ago? And how is it that these ancient Cambrian fossils have soft tissues that are so well preserved that it can be "dissected" and analyzed after more than 500 million years?

> *The early Cambrian Guanshan biota of eastern Yunnan, China, contains exceptionally preserved animals and algae. Most diverse and abundant are the arthropods, of which there are at least 11 species of trilobites represented by numerous specimens. Multiple specimens of both species contain the preserved remains of an expanded stomach region (a "crop") under the glabella, a structure which has not been observed in trilobites this old, despite numerous examples of trilobite gut traces from other Cambrian Lagerstätten. In addition, at least one specimen of Palaeolenus lantenoisi shows the preservation of an unusual combination of digestive structures: a crop and paired digestive glands along the alimentary tract. This combination of digestive structures has also never been observed in trilobites this old.... The presence and combination of these digestive features in the Guanshan trilobites contradicts current models of how the trilobite digestive system was structured and evolved over time.*[241]

Cambrian trilobites and anomalocaris had eyes, brains, and nervous systems without a clear evolutionary path. More amazingly, they had different eye types. It was as if they had suddenly appeared out of *"thin air."* The study of soft tissue and the preservation of Cretaceous proteins in dinosaur bones[242] continues to create a stir in the scientific literature and the news.

---

exceptionally preserved trilobites of the Guanshan Biota (Early Cambrian, China). PLoS One. 2017 Sep 21;12(9):e0184982.
241 Hopkins, M.J., F. Chen, S. Hu, and Z. Zhang. 2017. Ibid.
242 Schroeter, E.R., and others. 2017. Expansion for the *Brachylophosaurus canadensis*Collagen I Sequence and Additional Evidence of the Preservation of Cretaceous Protein. J. Proteome Res. 2017162920-932.

How did sophisticated animals with eyes, brains, digestive and nervous systems suddenly appear without predecessors? How did multiple differences in eyes appear together at the same time?

12 *Cyclic Global Flooding: the Rocks Testify*

The earth is utterly broken down, the earth is clean dissolved, the earth is moved exceedingly. The earth shall reel to and fro like a drunkard.[243] The waters of the flood were upon the earth... the same day were all the fountains of the great deep broken up, and the windows of heaven were opened. And the waters prevailed, and were increased greatly upon the earth... And the waters prevailed exceedingly upon the earth; and all the high hills, that [were] under the whole heaven, were covered. And all flesh died that moved upon the earth, both of fowl, and of cattle, and of beast, and of every creeping thing that creepeth upon the earth, and every man: All in whose nostrils [was] the breath of life, of all that [was] in the dry [land], died. [244]

High-energy, extraterrestrial triggering events, sufficient to cause global flooding, would leave a worldwide geologic footprint. These unique conditions and events would have to be powerful enough to thrust the invasion of seas not just across local regions like the Columbia Basin or even the Mediterranean Sea but across continents. The power of such an event would be staggering: powerful enough to cause seas to race across the continents. The invasion of flood waters and tsunamis, together with their transported sediments would rip up the land and leave mass graveyards within the sediment pile. Today, marine sediments and turbidites are being identified all around the globe

---

243 Isaiah 24:19, 20.
244 Genesis 7:10-23.

throughout the Phanerozoic (Cambrian to Present) rock record, standing as a testimonial of extensive global flooding.

Footprint of a Worldwide Flood: Multiple Lines of Evidence Confirm 6 Major Cycles of Flooding since the Cambrian Explosion of Life

Numerous lines of evidence point to 6 cycles of global catastrophic flooding since the early Cambrian. These were accompanied by an outpouring of CO_2 giving evidence of cyclic eruptions of volcanic magmas and gases into the oceans and atmosphere. These sediments were deposited, often times, as near continuous layers of sediments across vast spans of continents. Burial and preservation of Cambrian marine life was due, in most cases, to rapid fine grained mud flows that suffocated their victims, providing a unique means of preservation. Turbidites are typically triggered by earthquakes often resulting in rapid catastrophic burial.

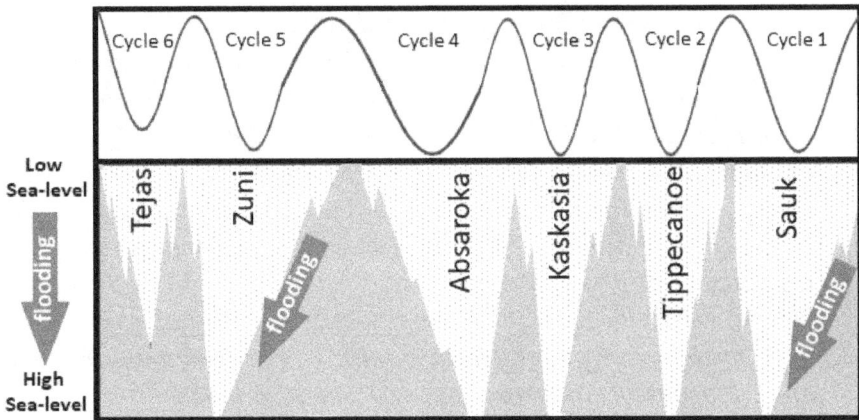

During just the past 12% of Earth's history, referred to as the Phanerozoic, the continental land masses have been invaded by the seas through these six major cycles of global flooding and erosion.[245] Do these actions and other signals point to another cycle of destruction? But then again, some question whether there

---

245 https://www.ck12.org/earth-science/paleozoic-and-mesozoic-seas/lesson/Paleozoic-and-Mesozoic-Seas-HS-ES/

is rock solid scientific evidence of past global flooding. The following provides the undeniable evidence:

First Line of Evidence: a Great Erosion Event at the Precambrian-Cambrian Boundary – the Great Unconformity

The first line of evidence is the incomparable global erosion[246] event at the Precambrian-Cambrian boundary, referred to as the *Great Unconformity*. Scientists believe that both glaciation and extensive flooding are the causes for this worldwide feature.

The Great Unconformity, a profound gap in Earth's stratigraphic record often evident below the base of the Cambrian system, has remained among the most enigmatic field observations in Earth science for over a century. While long associated directly or indirectly with the occurrence of the earliest complex animal fossils, a conclusive explanation for the formation and global extent of the Great Unconformity has remained elusive. Here we show that the Great Unconformity is associated with a set of large global oxygen and hafnium isotope excursions in magmatic zircon that suggest a late Neoproterozoic crustal erosion and sediment subduction event of <u>unprecedented scale</u>. These excursions, the Great Unconformity, preservational irregularities in the terrestrial bolide impact record, and the first-order pattern of Phanerozoic sedimentation can together be explained by spatially heterogeneous Neoproterozoic <u>glacial erosion totaling a global average of 3–5 vertical kilometers</u>, along with the subsequent thermal and isostatic consequences of this erosion for global continental freeboard. [247]

246 Keller, C.B. and others. 2018. Neoproterozoic glacial origin of the Great Unconformity. PNAS. Volume 116. Number 4.

247 Keller, C. B. and others. 2019. Neoproterozoic glacial origin of the Great Unconformity. PNAS 116 (4) 1136-1145. Freeboard: The average level of the sea surface compared to the Continents.

Second Line of Evidence: **Six Worldwide Transgressions and Regressions (Flooding) associated with Worldwide tectonism, High Sea Levels, and the Great Unconformity**

The second line of evidence is based upon 6 worldwide invasions of land by the seas referred to as the Sauk, Tippecanoe, Kaskasia, Abseroka, Zuni, and Tejas Transgressions. For example, the *Sauk Transgression* rests upon the Great Unconformity at the Grand Canyon:

> *The Sauk transgression was one of the most dramatic global marine transgressions in Earth history. It is recorded by deposition of predominantly Cambrian non-marine to shallow marine sheet sandstones unconformably above basement rocks far into the interiors of many continents. ... the marine transgression across a greater than 300-km-wide cratonic region took place during an interval 505 to 500 million years ago—more recently and more rapidly than previously thought. We redefine this onlap as the main Sauk transgression in the region. Mechanisms for this rapid flooding of the continent include thermal subsidence following the final breakup of Rodinia, combined with abrupt global eustatic changes driven by climate and/or mantle buoyancy modifications.* [248]

Third Line of Evidence: **High Sea Levels, low topography, and seawater displacement by upwelling mid-ocean ridge systems**

The third line of evidence is based on high sea levels that occurred concurrently with relatively low lying continental terrain. And these *high sea levels were further elevated by upwelling ocean ridges which displaced ocean water and contributed to the flooding of the continents*. As the molten magmas rise through the mantle and into the crust of the oceans, *their buoyancy raises crustal oceanic spreading centers[249] as high as mountains and*

---

248 Karlstrom, K. and Others. 2018. Cambrian Sauk transgression in the Grand Canyon region redefined by detrital zircons. Nature Geoscience. Volume 11, pages438–443
249 National Geographic. 2015. Seafloor spreading. nationalgeographic.org.

causes the displacement of ocean waters onto the continents.
Thousands of miles of mid-ocean volcanic chains were being lifted
simultaneously along the sea floor, all at once, as illustrated on the
back cover of this book. Their sheer volumes and the
accompanying frequency of earthquakes would repeatedly displace
water onto the continents. Does this upwelling of ocean floor
occur on a cyclical basis?

**Fourth Line of Evidence: Layer upon Layer of Continent-wide
stratigraphic facies with constant compositions and thicknesses**

The fourth line of evidence is the existence of numerous
sedimentary rock units spread across continental fragments of
Rodinia. By the Cambrian period, a relatively rapid breakup[250] of
Rodinia resulted in the development of extensive accumulation of
sediments along the fragmental continental margins of Gondwana,
Laurentia, Baltica, and Siberia. Flow patterns suggest that these
massive quantities of sediments were transported by surging walls
of water across the continents, again as evidenced by turbidite
beds. In the mid to late Cambrian, the Sauk transgression flooded
these continental landmasses with shallow seas in numerous pulses
which transported blankets of sand and muds across the North
American portion of Laurentia. Today extensive stratigraphic
units arrayed like a layer cake, cover enormous areas of the Earth's
surface. *The Tapeats sandstone covers much of North America
and is correlated with analogous geologic sequences in North
Africa, Eastern Greenland, Israel, South Australia, and portions
of South America.*

Beyond the Laurentian continental margin, huge quantities of thick
Belt Supergroup sediments accumulated along the shores of
Laurentia with more than 15 kilometers in thickness in some
places. The Belt Supergroup extends from Montana, Idaho,
Wyoming, Washington, and northward into Canada, where it is
referred to as the Purcell Supergroup. These sediments are derived
from erosion of the older crustal cores of Rodinia. Beneath the

---

250 Ross N. Mitchell, David A.D. Evans, and Taylor M. Kilian. 2010. Rapid
Early Cambrian rotation of Gondwana. Geology. 38(8):755-758.

Tapeats sandstone are found fragmented basement rocks testifying of turbulence and erosive flooding of invading seas.[251, 252]

Fifth Line of Evidence: **Sedimentation Rates may have been far faster than can be attributed to uniformitarianism, supporting rapid catastrophic flooding.**

Some studies indicate that the sedimentological rate, at least for certain rock layers, is far faster than the rates typically assigned to most stratigraphic units.

Based on the sedimentation analysis of the COS [Cambrian-Ordovician Sandstone Sequence] from the Leningrad district, "pure" sedimentation time for Lower Paleozoic sands can be estimated at 100–200 yr. The paradox is that geological time of the Sablinka sequence formation amounts to 10–20 Ma (Tugarova et al., 2001, p. 89). The authors explain this paradox by the rewashing of sediments in shallow water marine conditions with active lithodynamics, where processes of accumulation and seafloor erosion occur side by side and replace one another depending on parameters of storms and currents.[253]

Turbidite deposition, like that of the 1929 Newfoundland turbidite mud flow that cut the trans-Atlantic cables are found worldwide throughout the Rock Record. Many turbidite sequences blanket multiple continents consistent with global flooding episodes, much faster than uniformitarian age assignments might suggest. These turbidite deposits were laid down over vast distances in days and weeks; not thousands to millions of years as is typically assigned.

---

251 Bush, J.H., R.C. Thomas, and M.C. Pope. 2012. Sauk Megasequence Deposition in Northeastern Washington, Northern Idaho, and Western Montana. AAPG Memoir.

252 Alsalem, O.B. and others. 2017. Paleozoic sediment dispersal before and during the collision between Laurentia and Gondwana in the Fort Worth Basin, USA. Geosphere, Volume 14 (1): 325-342.

253 Berthault, G. and others. 2009. Reconstruction of Paleolithodynamic Formation Conditions of Cambrian–Ordovician Sandstones in the Northwestern Russian Platform. Lithology and Mineral Resources, Vol. 46, No. 1, pp. 60–70. © Pleiades Publishing, Inc.

Sixth Line of Evidence: **Enormous spike in Carbon Dioxide (CO2) and The Great American Carbonate Bank lining Interior Basins and Continental Platforms**

Is today's global warming a myth? The sixth line of evidence indicates that extreme levels of carbon dioxide in the atmosphere during the Cambrian was far in excess of today's levels.[254] These enormous quantities of CO_2 and hothouse temperatures are evidence of a catastrophic Cambrian event. *This Cambrian CO_2 spike attests to the enormity of worldwide volcanic eruptions.* Much of the volcanism was attenuated by the overlying seawater since many of the exploding volcanics released their gases and ash along thousands of miles of underwater volcanic chains under the ocean surface. These volcanic chains led to the release of large volumes of carbon dioxide driving the precipitation of large volumes of carbonates along continental shelves and inland seas.

In a new paper in the journal Science, three scientists including geologist Dennis Kent of Lamont-Doherty Earth Observatory, have found that the concentration of carbon dioxide in the air <u>doubled after each episode of volcanism.</u> All volcanoes spew CO_2, but these eruptions, lasting 20,000 years each and <u>dwarfing anything seen or imagined in human times</u>, must have spewed a lot.[255]

Carbon dioxide reacts with water to form bicarbonate ions and through a sequence of reactions, calcium and bicarbonate ions combine to form calcium carbonate (limestone) in the oceans.

When Iceland's Eyjafjallajokull volcano erupted in 2010, the fallout was global. Thousands of tons of carbon dioxide

---

254 Berner, R.A. and Z. Rothavala. 2001. GEOCARB III: A Revised Model of Atmospheric CO2 Over Phanerozoic Time. American Journal of Science, Volume 301, Pages 182-204.
255 Krajick, Kevin. 2011. Giant CO2 Eruptions in the Backyard? State of the Planet. Earth Institute. Columbia University. ei.columbia.edu. Based on Schaller, N.F. and others. 2011. Atmospheric Pco2 Perturbations Associated with the Central Atlantic Magmatic Province. Science. Volume 331, Issue 6023, pages 1404-1409.

were released and mineral ash shot 30,000 feet into the atmosphere, halting air travel across Europe. But closer to the source, scientists observed a more local phenomenon: a nearby river, the Hvanná, began to run milky white and large chalky, chunks formed along its banks. The strange substance, it turns out, was essentially solid CO_2—a carbonate, technically—released from the explosion and trapped in solid form rather than released as a gas.[256]

Seventh Line of Evidence: **Causative Factors: Wobbling and Tilting of Earth's Axis and the Reset of the Magnetic Field**

The seventh line of evidence is the fingerprint of the causative energy that served as the collective motivating force or forces that drove the world's oceans time and again over land for extended periods. At the very beginning of the Cambrian just such evidence is found, recorded in the rock record like a smoking gun.

Analysis of Vendian to Cambrian paleomagnetic data shows <u>anomalously fast rotations and latitudinal drift for all of the major continents</u>. These motions are consistent with an Early to Middle Cambrian inertial interchange true polar wander event, during which <u>Earth's lithosphere and mantle rotated about 90 degrees in response to an unstable distribution of the planet's moment of inertia</u>. The proposed event produces a longitudinally constrained Cambrian paleogeography and <u>accounts for rapid rates of continental motion during that time</u>. The new ages, along with <u>paleomagnetic data, indicate that continents moved at rapid rates that are difficult to reconcile with our present understanding of mantle dynamics</u>. We propose that rapid continental motions during the Cambrian period were driven by an interchange event in Earth's moment of inertia tensor. The age constraints on the geophysical data indicate that the rapid continental motions occurred during the same time

---

256 Ferry, David. 2016. Researchers In Iceland Are Showing Us How to Fix Carbon Emissions. outsideonline.com

interval as the Cambrian evolutionary diversification and therefore the two events may be related.[257]

Rapid active sea floor spreading and subduction of the sea floor readily serve as additional fingerprints of the causative event that drove water over the land. True paleo-wander evidence coupled with extensive volcanic activity along both island arc and ocean ridge systems[258] in the late Precambrian give testimony of the breakup of Rodinia and the opening of the Iapetus Ocean.[259]

Massive submarine volcanism, together with its associated phenomena of warming, sea-level rise, and widening of warm-weather zones, is proposed to be the chief extrinsic trigger for the Phanerozoic revolutions [e.g., Cambrian Explosion]. The later Mesozoic was characterized by continental rifting, which accompanied massive submarine volcanic eruptions that produced large quantities of nutrients and carbon dioxide. This activity began in the Late Triassic and peaked in the mid- to Late Cretaceous. The Early Cambrian was also a time of rifting and may likewise have been marked by large-scale submarine volcanism.[260]

Violent Earth wobble, rapid shifting of the Earth's tectonic plates, the breakup of Rodinia, and worldwide volcanism, all point to a seminal period in Earth history. Another fingerprint of this causative energy is the record of the Earth's magnetic field. Earth was at a critical point at a time when the dynamo almost

---

257 Kirschvink, Joseph L. and others. 1997. Evidence for a Large-Scale Reorganization of Early Cambrian Continental Masses by Inertial Interchange True Polar Wander. Science. Volume 277. Pages 541-545.
258 Beranek, L. and others. 2012. Tectonic Significance of Upper Cambrian–Middle Ordovician Mafic Volcanic Rocks on the Alexander Terrane, Saint Elias Mountains, Northwestern Canada. The Journal of Geology. Volume 120, Issue No. 3, pages 293-314.
259 Dunning, G.R. and others. 1991. Cambrian island arc in Iapetus: Geochronology and geochemistry of the Lake Ambrose volcanic belt, Newfoundland Appalachians. Geological Magazine 128(01):1 – 17.
260 Vermeij, Geerat, J. 1995. Economics, volcanoes, and Phanerozoic revolutions. Paleobiology. 21(2). Pages 125-152.

collapsed.[261]   Something had to kickstart the field.   Could the shaking of the Earth be a causative factor in restarting the Earth's magnetic field? Afterall, the breakup and rapid rotation of Rodinia occurred at the same time interval. Coincidence? More likely, *the wobbling of the Earth, the rapid movement of the continents, the rotation of the continents from the pole to the equator, and the restart of the Earth's magnetic field are at least partially a consequence of a common cyclical event.*

Eighth Line of Evidence: **Hothouse to Extreme Hothouse Temperatures and the Melting of Snowball Earth**

The eighth line of evidence involves hothouse to extreme Cambrian hothouse temperatures that are consistent with global flooding[262] that caused extensive glacial melting.   These high ocean surface water temperatures also coincide with the rotation of glacier covered landmasses from the pole to the equator accelerating melting of *Snowball Earth.   Glacial meltwaters* would provide a major contribution to the high sea levels.[263]   As testified by the evidence, sea level rise was driven by at least three synergistic factors. First, plates shifted towards the equator, and sea level rose due to *glacial melt waters* from the thawing of Snowball Earth.   Secondly, plate rotation was accompanied by volcanism at island arcs along continental shores which triggered *earthquakes and destructive tsunamis*.   And thirdly, the ocean floor rose in response to the upward *buoyant force of the magmas at oceanic volcanic ridge lineaments* causing the displacement of huge volumes of seawater that raced across the continental crust. Are core-mantle nuclear explosions cyclical? Is it possible that nuclear materials buildup, explode, scatter, and re-accumulate?

---

261 Saplakoglu, Yasemin. 2019. Earth's Magnetic Field Nearly Disappeared 565 Million Years Ago. Livescience.com. also see Bono, Richard K. and others. 2019. Young inner core inferred from Ediacaran ultra-low geomagnetic field intensity. *Nature Geoscience* volume 12, pages143–147.
262 Scotese, C.R. 2016. A New Global Temperature Curve For The Phanerozoic. GSA Annual Meeting in Denver, Colorado, USA - 2016, Volume: Geological Society of America Abstracts with Programs. Vol. 48, No. 7.
263 Hallam, A., 1992. Phanerozoic Sea-level. Columbia University Press, New York, 266 pages.

13 *Mass Graveyards: The Living Among the Dead*

All flesh that moved on the earth perished, birds and cattle and beasts and every swarming thing that swarms upon the earth, and all mankind; of all that was on the dry land, all in whose nostrils was the breath of the spirit of life, died.[264] By the breath of God ice is given, and the broad waters are frozen fast.[265] The waters become hard like stone, and the face of the deep is frozen.[266]

Sudden, unrelenting shock waves caused by a gargantuan shift of the continental shelf, sent fine mud cascading down the coastal slopes all around the globe, descending like great underwater clouds upon the offshore habitat of soft bodied life forms. All along the deep off-shore continental shelves, terror reigned in the once quiescent deep waters. Bottom feeders tried to scurry away in vain; there was no way of escape. Within seconds there was a shift from light to darkness and even the creatures with their compound eyes could see nothing but gloom. Anomalocaris, trilobites, jellyfish, gastropods, annelids, chordates, conodonts, nautilus cephalopods, brachiopods, echinoderms, sponges, mollusks, and representatives of numerous other phyla were instantly sealed alive in muddy tombs.

The sudden destruction left its deadly toll of massive graveyards. Their fossilized bodies are found entombed all around the once offshore fragments of Rodinia from Chengjiang (China), Yakutia (Siberia), Newfoundland, Pennsylvania, Emu Bay (Australia),

---

264 Genesis 7:21, 22.
265 Job 37:10.
266 Job 38:30.

Morocco, Sweden, Greenland and all along the west coast of North America from British Columbia to Mexico to name but a few. And in those moments, as the continents violently vibrated and trembled, volcanic eruptions released a fury of magma, ash, and gases, causing layer upon layer of mudstones to be deposited all along the continental shelf. A few escaped, like the velvet worm and the nautilus that are still living today, passing their precious genetic code to their descendants. *The entire world was in upheaval.*

The continent of Laurentia, like the other fragments of Rodinia had been ground by glaciers. Incredible forces tore Laurentia apart, forming the St. Lawrence Seaway in what geologists call a rift zone, much like the East African rift zone of today. But the shattering of the proto United States didn't stop there. The fracture at the St. Lawrence Seaway can be traced to an older scar known as the Mid-Continental Rift System (MCRS) that opened up through Michigan. Southward from Michigan, there's the Oklahoma rift zone (aulacogen), and the New Madrid rift zone (aulacogen) extending south to the Mississippian embayment. And all along this Laurentian zone of the shattered planet, magma from the great deep gushed upwards to the surface colliding with invading seawaters. *A prominent conductor traces out Cambrian rifting in Arkansas, Missouri, Tennessee, and Kentucky; this linear conductor has not been imaged before and suggests that the Cambrian rift system may have been more extensive than previously thought.[267]*

A bonanza of ancient sea-life is found all along the margins of Rodinia's continental fragments and in the rift valleys in the interiors of these fragments. Along the ancient St. Lawrence rift zone, numerous sea creatures found their way to the interior of North America. Along the shores of Lake Ontario's Chaumont Bay, I've found thousands of cephalopods buried abruptly in the lime muds, much like those found in the Cambrian fossil beds of Morocco.

---

267 Bredosian, Paul A. 2016. Making it and breaking it in the Midwest: Continental assembly and rifting from modeling of EarthScope magnetotelluric data. Precambrian Research. Volume 278.

Cambrian Rift Zone of Morocco: Transitional Fossilized Ecosystem or Deep Geological Time?

The surface of the Earth began to shake wildly, as chaos raced through the swampland home of thousands of animals. In the distance, billowing clouds of volcanic ash poured ever upward into the atmosphere along a front, tens of thousands of miles long. And as many as six major distinct surges of magma[268] jetted out of the Earth in a great fury. The ash soon blocked out the Sun, reducing it to a strange red glow, causing an eerie darkness that rapidly filled the once blue sky. Unbeknownst to its victims, a giant chasm was opening up along the edge of the West African Craton (WAC) in what scientists today call the "Cambrian Rift."[269]

Magmas rising to the surface from the primitive mantle sources caused the surface of the Earth to rip, fracture, and fault along the underwater carbonate shelf. Volcanoes spit out their ash flows to the air that were accompanied by mudflows that were sent racing along the seafloor.[270] A treasure trove of fossil beds[271] in what is modern day Morocco resulted from rapid burial of these offshore creatures. In the Moroccan site, referred to as the Fezouata formation of the Ordovician period (in the strata just above the Cambrian layers), over 160 marine genera have been found.[272] Morocco provides a wide collection of fossil graveyards ranging from offshore cephalopods, numerous trilobites, echinoderms, and

---

268 Pouclet, A and others. 2018. Review of the Cambrian volcanic activity in Morocco: geochemical fingerprints and geotectonic implications for the rifting of West Gondwana. International Journal of Earth Sciences. Volume 107, Issue 6, pp 2101–2123.
269 Berrada, H. and others. 2018. Example of post-Ediacarian complex volcanism emplaced during the Cambrian rifting in the Western High-Atlas, Morocco: Geochemical study and geotectonic significance. Bulletin de l'Institut Scientifique, Rabat, Section Sciences de la Terre, 2018, Number 40, 115-130.
270 Berrada and others, 2018. Ibid. Figure 14.
271 http://nautiloid.net/fossils/sites/alnif/nautiloid.html
272 Van Roy, Peter and others. 2015. The Fezouata fossils of Morocco; an extraordinary record of marine life in the Early Ordovician. Journal of the Geological Society. 172(5):541. Also see Geological Society of London. "Spectacular Moroccan fossils redefine evolutionary timelines." ScienceDaily. ScienceDaily, 7 July 2015.

horseshoe crab-like creatures (genus Illaenus). And scientists have learned that the Moroccan Ampyx trilobites were also suddenly buried in-place in strange long lines[273] as if they were an infantry of soldiers marching to battle. Burial of the trilobites in this learned group behavior not only underscores that these animals seemingly appeared instantaneously as a part of the Cambrian Explosion but they already had community roles just as bees in hives seen today. How could such a sense of community develop so suddenly?

Importantly, numerous fossils from the Fezouata formation have preserved soft body parts in worm like creatures[274] and trilobites.[275] *All this has led some researchers to believe that the Ordovician diversification is a continuation of the Cambrian Explosion! Or are there other possible explanations? Could we be viewing the fossilization of animals living at the same time but in different environmental zones (Transitional Ecosystem Flood Stratigraphy)?* Maybe. You be the judge!

Continuing further inland, beyond the Moroccan Atlas Mountains, one finds fossils that lived near shore. And many of these creatures hovered over the sea floor genera. These upper marine dwellers include plesiosaurs (genus Thililua), placoderms, prawns, ray-finned fishes, sea turtles, and sharks among numerous other marine creatures.

And these fossils transition inland to the ancient swamplands of Northern Africa just like they would in a modern terrain of increasing elevation. And these swamplands and their swamp creatures have been buried under the great Sahara desert until recent scientific expeditions have uncovered their secrets.

---

273 Vannier, J. and others. 2019. Collective behaviour in 480-million-year-old trilobite arthropods from Morocco. Scientific Reports **volume 9**, Article number: 14941.

274 Moskowitz, C. January 2008. Fossil of Extinct Armored Worm Discovered. livescience.com.

275 Guitierrez-Marco, J.C. and others. 2017. Digestive and appendicular soft-parts, with behavioural implications, in a large Ordovician trilobite from the Fezouata Lagerstätte, Morocco. Scientific Reports 7, Article number: 39728

The Swamps and Shores of North Africa

Legend has it that the Sahara desert was once underlain by giant lake Triton. According to several accounts:

Stories of Lake Triton's demise in Africa... this lake disappeared in a catastrophic earthquake that occurred in Africa wherein this lake broke into the ocean.[276]

The Central Mountains of North Africa were surrounded by a conglomeration of lakes, big marshes where Atlas cedars, wild fig trees, ash trees, lime trees, and willows grew. On the banks, roses and papyrus flourished. The waters were the habitat of crocodiles, hippopotami and tortoises. Elephants, rhinoceros, buffalo, wild oxen and great antelopes found a favourable habitat within this paradisiacal area.[277]

A tremendous earthquake along a fault broke open the Atlas Mountains. In one terrible night, as the Earth's crust shivered violently and erupting volcanoes thrust deadly rocks and boiling lava into the atmosphere, all the water emptied from the lake. Torrents of foaming liquid, combined with volcanic debris, completely buried 200 cities built around the Triton Sea.[278]

Recent analyses of the Sahara, using data from NASA's Shuttle Radar Topography Mission, has led to the discovery of a former enormous lake covering upwards of 42,000 square miles. The lake boundary was corroborated by Paleolithic settlements bordering the lake.[279] And interestingly, paleontologists are finding fossil

---

276 Pilotte, R. 2019. Flood Legends: Global from Local and Some Evidence of Each. Trafford Publishing. Pages 324. ISBN: 9781490795652
277 Zitman, W.H. 2006. Egypt: Image of Heaven: The Planisphere and the Lost Civilization. Adventures Unlimited Press. Pages 311.
278 Andrews, Shirley. 2018. Atlantis: Insights from a Lost Civilization. Authorhouse. Pages 292.
279 Earth Science, Research News, Science & Nature. 10 December 2010. Ancient Megalake discovered beneath Sahara Desert. Smithsonian Insider.

remains of humans, sauropods, whales, spinosaurus, crocodiles, and catfish, not to mention snakes and turtles.[280]

Amazingly, these mass burial sites transition from the offshore burial of soft-bodied creatures to the swamplands discovered under the sands of the Sahara. Could these mass burials represent a __continuous sequence of ecosystems that were all involved in the same catastrophic event? Are transitional ecosystem graveyards found elsewhere?__

Mystery of the Fossilized Whales

Shock waves of devastating proportions must have caused the Earth to *reel like a drunkard* through time. Death reigned in the Pacific Ocean near the center of the impact where tranquility once ruled. Enormous whales, fish, and all marine life in the vicinity of the supersonic blast of the exploding fireball were thrown with sea water hundreds of feet into the air. *It was Nature's form of blast fishing.*[281] In an instant, massive walls of water carried the dead to the coast line of what would become known today as Chile and Peru. Wave upon wave crashed hundreds of feet inland depositing the dead like toothpicks. In Chile, at Cerro Ballena, perhaps the dead were cast upon the desert shores by a series of storm waves:

> *Identified in the beds were over 40 individual rorquals - the type of large cetacean that includes the modern blue, fin and minke whales. Among them were other important marine predators and grazers. "We found extinct creatures such as walrus whales - dolphins that evolved a walrus-like face. And then there were these bizarre aquatic sloths," recalls Nicholas Pyenson, a palaeontologist at the Smithsonian's National Museum of Natural History...The team immediately noticed that the skeletons were nearly all complete, and that their death poses had clear commonalities. Many had come to rest facing in the same direction and upside down... The*

---

280 Soders, Erich. 2019. 10 Amazing Fossils Found in the Sahara Desert. listverse.com.
281 Galbraith, Kate. February 4, 2015. The Horrors of Fishing with Dynamite. The New York Times.

> *researchers believe the configuration of the coastline at Cerro Ballena in the late Miocene Epoch worked to funnel carcases into a restricted area where they were lifted on to sand flats just above high tide, perhaps by storm waves.*[282]

The setting was much like that of Alaska's inlet at Lituya Bay where the world's highest tsunami wave was recorded at an elevation of 1720 feet above sea level. Is it possible, that the dead marine animals of Cerro Ballena were likewise carried inland and buried, layer upon layer, by crashing tsunami waves of extraordinary height and power? Elsewhere, scientists have found large fossil beds in the Pisco formation in Peru that contain hundreds of whale fossils that appear to have been flash deposited along with other marine and land animals including fish, turtles, seals, porpoises, and penguins together with ground sloths.

> *The well-preserved whales indicate rapid burial. The 346 whales within ~1.5 km² of surveyed surface... were distributed uninterrupted through an 80-m-thick sedimentary section. The diatomaceous sediment lacks repeating primary laminations, but <u>instead is mostly massive, with irregular laminations and speckles.</u>*[283]

Whale fossils have been found worldwide from Egypt[284] to the United States. Whales are found along the East Coast of the United States from Vermont[285] (more than 600 feet above sea level) southward to Georgia and Florida. What could have caused such towering waves of destruction? Researchers are discovering mega-tsunami evidence throughout the South Pacific region from Antarctica, Chile, and New Zealand.

---

282 Amos, Jonathon. 2014. Chile's stunning fossil whale graveyard explained. BBC News. bbc.com.

283 Brand, L.R. 2004. Fossil whale preservation implies high diatom accumulation rate in the Miocene–Pliocene Pisco Formation of Peru. Geology (2004) 32 (2): 165-168.

284 Groves, D. 2016. Huge prehistoric whales found in Egyptian desert. https://us.whales.org/2016/01/21/.

285 Fessenden, Marissa. June 2016. Landfill Surprises Scientists with 12-Million-Year-Old Whale Fossils. Smithsonian.com.
also Wiertlieb, M. 2014. A Whale In Vermont? The Story Behind The State's Most Famous Fossil. vpr.org.

Large asteroid impacts are rare, and those into the deep ocean are rarer still. The Eltanin asteroid impact... occurred at a time of great climatic and geological change associated with the Pliocene–Pleistocene boundary. <u>Numerical models of the event indicate that a megatsunami was generated, although there is debate concerning its magnitude and the region-wide extent of its influence.</u> We summarise the existing evidence for possible Eltanin megatsunami deposits in Antarctica, Chile and New Zealand, while also examining other potential sites from several locations, mainly around the South Pacific region. In reviewing these data we note that these events were unfolding at the same time as those associated with the Pliocene–Pleistocene boundary and, as such... it raises interesting questions about the role potentially played by such catastrophic events in contributing to or <u>even triggering epochal transitions.</u>[286]

Evidence of the Eltanin asteroid impact includes a strong iridium[287] signature in the sediments.

Evidences of a Flooding across Western Europe and the Mediterranean Coasts

Joseph Prestwich presented a stunning memoir to the Royal Society in 1893[288] which claimed, based on stratigraphic and paleontological evidence, that *much of Western Europe and the Mediterranean coasts were submerged in a glacial or post-glacial period.* Perhaps most stunning is his list of animal bones found in fissures of high ridges well above sea level, including the remains

---

286 Goff, J. and others. 2012. The Eltanin asteroid impact: Possible South Pacific palaeomegatsunami footprint and potential implications for the Pliocene-Pleistocene transition. Journal of Quaternary Science 27(7):660-670.
287 Kyte, F. 2002. Iridium concentrations and abundances of meteoritic ejecta from the Eltanin impact in sediment cores from *Polarstern* expedition ANT XII/4. Deep Sea Research Part II: Topical Studies in Oceanography. Volume 49, Issue 6, 2002, Pages 1049-1061.
288 Prestwich, Joseph. 1893. On the evidences of a submergence of Western Europe, and of the Mediterranean coasts, at the close of the glacial or so-called post-glacial period, and immediately preceding the neolithic or recent period. Philosophical Transactions of the Royal Society. Pages 80-89.

of *"felis, lynx, wolf, hyaena, bear, lagomys, hare, mammoth, rhinoceros, wild boar, horse, ox, deer, antelope together with land shells of various living species."* Prestwich draws the following conclusion:

> *The bones are mostly broken and splintered into innumerable sharp fragments, and evidently are not those of animals devoured by beasts of prey; nor have they been broken by man. It is not possible to suppose that animals of such different natures, and having such different habitats, could in life ever have herded together. Difficult as the alternative is, the author sees no other explanation of the phenomena than that of a <u>wide-spread temporary submergence, accompanied by strong earth tremors</u>. In such a case it is easy to conceive that as the waters gradually advanced over the low lands, the animals of the plains would naturally seek safety on the higher grounds and hills. Flying in terror, and cowed by the common danger, the Ruminants and other Herbivores, together with the Carnivores, would, as in the case of the flooding in our days of large deltas, alike seek refuge on the same safety spot. Where that spot was an isolated hill, they would, if it were not out of reach of the flood waters, eventually suffer the same fate. Subsequently the detached limbs and bones, carried, together with the surface debris, by the effluent currents into the open fissures, were subjected to the clashing of the rubble and the fall of large fragments of rock from the sides of the fissures, and were crushed and broken in the way they are always found.*

Many fossil beds around the world include the remains of predators and prey, herded together by raging flood waters. In several instances the animals appear to have been driven together from areas of contrasting climates such as the Maryland's Cumberland Cave[289, 290] and the Norfolk forest-beds of England.

---

289 Gidley, J. W. 1913b. Preliminary report on a recently discovered Pleistocene cave deposit near Cumberland, Maryland. Proceedings of the United States National Museum 46:93–102.

Elsewhere, significant fossil deposits have been found in Chile, China, Canada, Argentina, Montana, Wyoming to name a few. An ichthyosaur graveyard demonstrated the treacherous nature of the flood waters which carried its victims out to sea and buried them in clouds of turbid mud:

> *According to the German-Chilean research team, the fish-lizard lived and hunted along the northeastern edge of a deep sea that then separated the Antarctic continent from Patagonia. Adults and juveniles hunted in groups in an underwater canyon rich with squid and small fish, their most important prey. As the continent gradually broke apart, earthquakes or avalanches on the steep slope occasionally unleashed devastating mudflows that sucked everything in their path down with them, including the marine reptiles. "The air-breathing fish-lizards became disoriented in the turbidity currents. They were sucked down hundreds of metres into the deep ocean," says Prof. Stinnesbeck. "The fine sediment that was swept along immediately entombed the dead or dying animals."[291]*

Numerous waterbodies were isolated by the rapid rise of the Rocky Mountains including Fossil Lake, Lake Uinta, and Lake Goshute. These waterbodies cover a great stretch of land including Wyoming, Utah, and Colorado. The Green River formation is classified as a *Lagerstätten* because of its wide and diverse assemblage of plant and animal fossils. Fossils range from stromatolites which are found in Precambrian rock units and still thrive in the world today. The region is famous for its fish fossils including a gar fish, ostracods, stingrays, bats, plants, crocodiles, sycamore trees, other invertebrates, vertebrates, and even primates.

---

290 Nicholas, Brother G. 1953. Recent Paleontological Discoveries from Cumberland Bone Cave. American Association for the Advancement of Science. The Scientific Monthly. Vol. 76. Pages 301-305.
291 Heidelberg University. 2014. One of the world's most significant finds of marine reptile fossils from the cretaceous period. Phys.org. Wolfgang Stinnesbeck, and others. 2014. "A Lower Cretaceous ichthyosaur graveyard in deep marine slope channel deposits at Torres del Paine National Park, southern Chile." *Geological Society of America Bulletin* (published online 22 May 2014).

*The fossils are trapped in **micrite beds, which are beds of very fine (1-5 microns) carbonate crystals (either calcite or aragonite). Micrite can precipitate from seawater or form from the breakdown of larger carbonate grains. There are two hypotheses concerning the depth and conditions of the lakes. The first is that the lake was deep and stratified (Bradley, 1948), the second is that the lakes were unstratified, and had varying depths and salt concentrations (Buchheim and Surdham, 1981).*[292]

On August 17, 1909, nearby at what is now known as Dinosaur National Monument, Earl Douglass, a paleontologist with the Carnegie Institute in Pittsburgh, Pennsylvania, found eight dinosaur tailbones in a sandstone outcrop. If we could roll back time, it would truly have been a real live Jurassic Park. Allosaurus and Deinonychus (related to the Velociraptors featured in the movie "Jurassic Park") roamed with herds of herbivores like Stegosaurus, Apatosaurus, and Diplodicus. All are buried in the sediments of sand and muds. The dinosaurs roamed the region when the continent was near the equator. Scientists are still debating the fate of the dinosaurs found at Dinosaur National Monument.[293] Perhaps they were buried by flood waters just like those at the neighboring **Standing Rock Hadrosaur Site** and those of the **Hell Creek Formation.**

The incredibly high density fossil rich localities of the Hell Creek Formation which spans parts of Montana, North Dakota, and South Dakota is the same formation where Dr. Mary Schweitzer and her team famously claimed to have found soft tissues in a T-Rex dinosaur.[294] However, over the years her findings have been

---

292 Ucmp.berkeley.edu/tertiary/greenriver. Also Bradley, W.H. 1948. "Liminology and the Eocene lakes of the Rocky Mountain Region." Geological Society of America Bulletin, 59: 635-648 And Buchheim, H.P. and Surdham, R. 1981 Paleoenvironments and fossil fishes of the Laney Member, Green River Formation, Wyoming In: J. Gray, A.J. Boucot and W.B.N. Berry, Communities of the Past. Hutchinson Ross, New York 425-52

293 Black, Riley. 2015. What Killed the Dinosaurs in Utah's Giant Jurassic Death Pit? Smithsonian.com.

294 Schweitzer, Mary H. and others. 1997. Preservation of Biomolecules in Cancellous Bone of Tyrannosaurus rex. Journal of Vertebrate Paleontology Vol. 17, No. 2. Pages 349-359.

challenged. Recent scientific evaluations seem to exonerate Schweitzer. According to a research team at Yale University, the original proteins have been chemically transformed into polymer compounds through glycoxidation and lipoxidation. [295]

> *The study offers some vindication for Mary Schweitzer, a molecular paleontologist at North Carolina State University, who has long contended that dinosaur soft-tissue samples that she and others have described are in fact endogenous material and not bacterial contaminants. "When you want to change the mindset of an entire discipline, it takes time. If this new study is the turning point, then I'll be really happy," Schweitzer says.[296]*

Recently, a team of scientists led by paleontologist Robert DePalma,[297] have made another exciting discovery within the Hell Creek formation. *As soon as DePalma started digging he noticed grayish-white specks in the layers which looked like grains of sand but which, under a hand lens, proved to be tiny spheres and elongated droplets.* In fact, he found microtektites by the millions; proof of a meteoritic impact. Astonishingly, his field finding validated that the KT impact event led to rapid entombment of a broad spectrum of plant and animal life in a single deposit. *All of it was quickly entombed and preserved in the muck: dying and dead creatures, both marine and freshwater; plants, seeds, tree trunks, roots, cones, pine needles, flowers, and pollen; shells, bones, teeth, and eggs; tektites, shocked minerals, tiny diamonds, iridium-laden dust, ash, charcoal, and amber-smeared wood. As the sediments settled, blobs of glass rained into the mud, the largest first, then finer and finer bits, until grains sifted down like snow.* In my opinion, DePalma had found the moment that led to the extinction of many of the world's creatures. Creatures from a broad range of transitional ecosystems were buried simultaneously.

---

295 Wiemann, J. and others. 2018. Fossilization transforms vertebrate hard tissue proteins into N-heterocyclic polymers. Nature Communications 9, Article Number 4741.

296 Morton, Mary C, 2019. Dinosaur soft tissues preserved as polymers. Earthmagazine.org.

297 Preston, Douglas. 2019. The Day the Dinosaurs Died. The New Yorker.

If, on a certain evening about sixty-six million years ago, you had stood somewhere in North America and looked up at the sky, you would have soon made out what appeared to be a star. If you watched for an hour or two, the star would have seemed to grow in brightness, although it barely moved. That's because it was not a star but an asteroid, and it was headed directly for Earth at about forty-five thousand miles an hour. Sixty hours later, the asteroid hit. The air in front was compressed and violently heated, and it blasted a hole through the atmosphere, generating a supersonic shock wave. The asteroid struck a shallow sea where the Yucatán peninsula is today. In that moment, the Cretaceous period ended and the Paleogene period began.

Some of the ejecta escaped Earth's gravitational pull and went into irregular orbits around the sun. Over millions of years, bits of it found their way to other planets and moons in the solar system. Mars was eventually strewn with the debris—just as pieces of Mars, knocked aloft by ancient asteroid impacts, have been found on Earth. A 2013 study in the journal Astrobiology estimated that tens of thousands of pounds of impact rubble may have landed on Titan, a moon of Saturn, and on Europa and Callisto, which orbit Jupiter—three satellites that scientists believe may have promising habitats for life. Mathematical models indicate that at least some of this vagabond debris still harbored living microbes. The asteroid may have sown life throughout the solar system, even as it ravaged life on Earth.

The asteroid was vaporized on impact. Its substance, mingling with vaporized Earth rock, formed a fiery plume, which reached halfway to the moon before collapsing in a pillar of incandescent dust. Computer models suggest that the atmosphere within fifteen hundred miles of ground zero became red hot from the debris storm, triggering gigantic forest fires. As the Earth rotated, the airborne material converged at the opposite side of the planet, where it fell and set fire to the entire Indian subcontinent. Measurements of the layer of ash and soot that eventually coated the Earth indicate that fires consumed about seventy per cent of the world's forests. Meanwhile, giant tsunamis resulting from the impact churned across the Gulf of Mexico, tearing up coastlines, sometimes peeling up hundreds of feet of rock, pushing debris inland and then sucking it back out into deep water, leaving jumbled deposits that oilmen sometimes encounter in the course of deep-sea drilling.[298]

Based on years of research, DePalma is convinced that the site had been created by an impact flood. *By the time the site flooded, the surrounding forest was already on fire, given the abundance of charcoal, charred wood, and amber he'd found at the site. The water arrived not as a curling wave but as a powerful, roiling rise, packed with disoriented fish and plant and animal debris, which, DePalma hypothesized, were laid down as the water slowed and receded.* The asteroid had initiated a shock induced tsunami that buried life brutally and rapidly as if the earth was moved by an incredible explosion. It was the Chicxulub disaster!

---

298 Preston, Douglas. 2019. IBID.

Living Fossils: Creatures that Defy Deep Time

In a flash, in an instant in time, from a geological perspective, animal life appeared on Earth. According to most scientists today, life exploded during the Cambrian; not 5,500 years ago but 550,000,000 years ago; suddenly appearing in an advanced state of evolution. In that moment, the vast majority of animal phyla that exist today, mysteriously appeared. Could life have started just 5,500 years ago? What is most astonishing and intriguing is the evidence that *numerous animal species that are living today can be dated all the way back to the Cambrian. How could these species have survived for as long as 550,000,000 years* during such a long period of ocean invasions of land and mass extinctions*? Among these are stromatolites,[299] sponges, the nautilus squid, jellyfish,[300] crinoids, velvet worms, and snails to name a few. Numerous other species, including the coelacanth, are said to have been swimming in the oceans for more than 360,000,000. How is this possible?* Is our assessment of geologic time inaccurate?

> *The primitive-looking coelacanth (pronounced SEEL-uh-kanth) was thought to have gone extinct with the dinosaurs 65 million years ago. But its discovery in 1938 by a South African museum curator on a local fishing trawler fascinated the world...[301]*

How is it possible that the soft tissues of animals, now fossilized in Cambrian deposits in China, Canada, Siberia and elsewhere, have been so delicately preserved? How is it that numerous animals that are found entombed in Cambrian rock units have persisted for 550,000,000 without significant change? Could it be that what we see in the rock column is really *Transitional Ecosystem Flood Stratigraphy*? You be the judge.

---

299 Southgate, P.N. 1980. Cambrian stromatolitic phosphorites from the Georgina Basin, Australia. Nature 285. Pages 395-397.
300 Cartwright, P. and others. 2007. Exceptionally Preserved Jellyfishes from the Middle Cambrian. PLoS ONE.
301 National Geographic. Coelacanth. On-line Nationalgeographic. Jan. 2020.

RECONSTRUCTING EARTH'S APOCALYPTIC PATTERN

And the stars of heaven fell unto the earth... and every mountain and island were moved out of their places...[302] and there followed hail and fire mingled with blood, and they were cast upon the earth...and as it were a great mountain burning with fire was cast into the sea: and the third part of the sea became blood; and the third part of the creatures which were in the sea, and had life, died and there fell a great star from heaven, burning as... a lamp, and it fell upon the third part of the rivers, and upon the fountains of waters...[303]

What is the underlying cause of the incredibly near-uniform cyclical nature of flooding? Is Cyclical Flooding the key to the framework of an Apocalyptic Pattern? By assembling the pieces of the pattern, we can determine if the pattern is repeated. If it is, we can confirm that it is not only the key for unlocking the mystery of the Cambrian Explosion, but also for unlocking the coming Apocalypse. The stage is being set to resolve the age old controversy sparked by Lyell and Darwin. Could Newton be right after all?

---

302 Revelation 6:13-15.
303 Revelation 8:7-11.

14 *Cambrian Apocalypse: 1ˢᵗ Cycle of Things to Come*

[Seek him] that maketh the seven stars and Orion, and turneth the shadow of death into the morning, and maketh the day dark with night: that calleth for the waters of the sea, and poureth them out upon the face of the earth...[304] Raging waves of the sea, foaming out their own shame...[305] then shall be great tribulation, such as was not since the beginning of the world to this time, no, nor ever shall be.[306]

Cambrian events establish a pattern of things to come. Samples that I had collected, along with oriented samples collected by French, were analyzed to determine pole directions with the objective of evaluating the early opening and closing of the Atlantic.[307] Analyses completed by French confirmed, along with other studies, that the *"Wilson Cycle" of the repeated opening and closing of ocean basins along old orogenic belts is a key process in the assembly and breakup of supercontinents.[308]* This chapter flow-charts key synergies associated with Global Flooding. With the 6 primary flood cycles as the backbone of the pattern, collateral events are shown to be interwoven with the pattern. But can we confirm the pattern with scientific evidence associated with multiple cycles?

---

304 Amos 5:8.
305 Jude 1:13.
306 Matthew 24:21.
307 R.B. French, D.H. Alexander, and R. Van der Voo. 1977. Paleomagnetism of upper Precambrian to lower Paleozoic intrusive rocks from Colorado. GSA Bulletin Volume 88 (12). Pages 1785-1792.
308 Wilson, R.W. et. Al., 2019. Fifty years of the Wilson Cycle concept in plate tectonics: an overview. Geological Society, London, Special Publications 470.

Cambrian Apocalypse Interdependent Events

The flow diagram provides a qualitative sequence, based on the scientific literature, illustrating the incredible convergence of numerous momentous events, interconnected with one another and the **Cambrian Explosion** (not necessarily in the following order):

- Snowball Earth
- Glacial Erosion (high $^{87}Sr/^{86}Sr$)
- Photosynthesis
- Great Oxygen Event
- Great Unconformity
- Multiple major Cosmic Events
- And/or Cyclic Nuclear Events
- Magnetic Field restart
- Wobble of Earth's axis
- Planet tilt
- Cambrian Explosion
- Deep ocean canyon formation
- Asteroid triggered Plate tectonics
- Water Jets
- Ocean Ridge uplift and water displacement to continents
- Supercritical fluid injection (décollement)
- Continental Rifting and breakup; even in the central U.S.
- Folding, faulting and Mountain Range formation
- Plate rotation
- Worldwide volcanism
- Massive CO_2 emissions (hothouse Earth)
- Sea water temperature rise
- Global Phosphogenic event
- Sauk Transgression
- Carbonate Platforms
- Turbidites
- Continental inundation by Flooding **and much more**

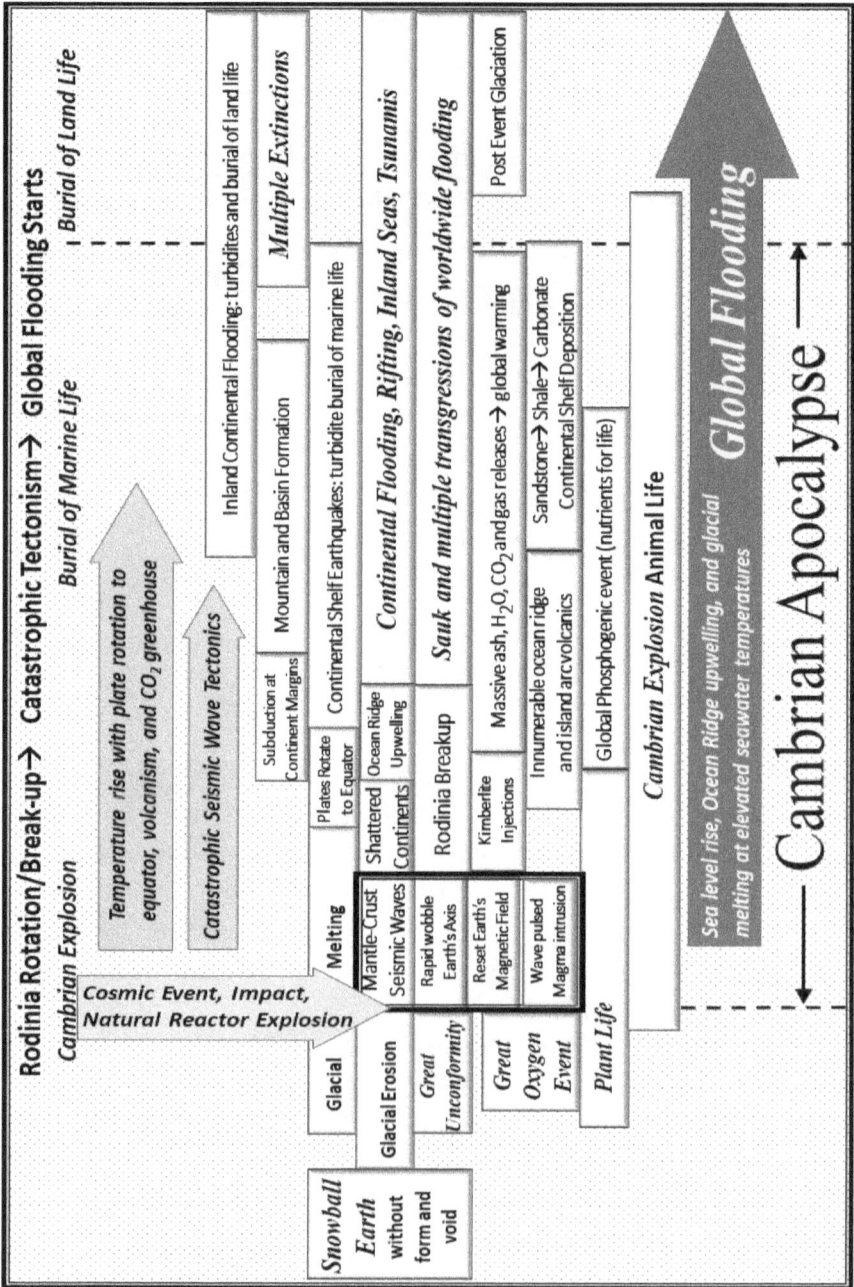

Snowball Earth: Erosion and the Great Unconformity

Prior to the Cambrian, it is believed that the surface of the planet was largely frozen. *Scientists believe that ... severe ice ages, which froze nearly or all of the surface, occurred between 750 million and 580 million years ago... "Once the polar oceans began to freeze, more sunlight was reflected off the white surfaces and cooling was amplified."*[309] The Great Unconformity across the globe is direct evidence of this intense glacial weathering and erosion. *Evidence of increased continental weathering during the Cambrian summarized by Peters and Gaines (2012), includes high $^{87}Sr/^{86}Sr$ and very negative ε_{Nd} seawater values.*[310] Glaciers grind down mountains and level the landscape. Waters at the base of glaciers enter cracks in the rock and upon freezing, break down the rock into pieces, whence raging rivers under the glaciers carry the broken fragments to the valleys below, and ultimately to the sea. Likewise, raging waters carrying everything from immense boulders, sand, and fine sized clays and silts, serving as the abrasive media, pummel and carve bedrock by acting as enormous belt grinders. The Great Unconformity at the Grand Canyon is a vast glacial and flood erosion plane in the rock record that produced 3-5 vertical kilometers of sediment:

> *The Great Unconformity, a profound gap in Earth's stratigraphic record often evident below the base of the Cambrian system, has remained among the most enigmatic field observations in Earth science for over a century... Here we show that the Great Unconformity is associated with a set of large global oxygen and hafnium isotope excursions in magmatic zircon that suggest a late Neoproterozoic crustal erosion and sediment subduction event of*

309 Ross, R. 2019. Snowball Earth: When the Blue Planet Went White. Livescience.com

310 McKenzie, R.N. and others. 2014. Plate tectonic influences on Neoproterozoic–early Paleozoic climate and animal evolution. Geology. Volume 42. Pages 127-130.

unprecedented scale. These excursions, the Great Unconformity, preservational irregularities in the terrestrial bolide impact record, and the first-order pattern of Phanerozoic sedimentation can together be explained by spatially heterogeneous Neoproterozoic <u>glacial erosion</u> totaling a global average of <u>3–5 vertical kilometers</u>, along with the subsequent thermal and isostatic consequences of this erosion for global continental freeboard.[311]

Mars, Enceladus: Fountains of the Great Deep

Recent scientific evidence[312] from Martian studies provides confirmatory evidence of the validity of the outpouring of oceans of water onto the surface of a planet by liquefaction (like squeezing the water out of a sponge) associated with an impact.

The Shallow Radar on Mars Reconnaissance Orbiter can image through … deposits, revealing their internal structure and composition. We find that these deposits are very rich in ice, which lies in horizontal slabs alternated with sand. The occurrence and volume of ice slabs increase toward the north pole. This ice may be the leftover of former ice caps that diminished during warm periods and therefore represent an important record of past Martian climate. The large volume of ice preserved within the… unit represents one of the largest water reservoirs on the planet.

Sounds a lot like a *Snowball* Earth. *Is the barren Martian landscape, of our past destined for our future?* The discovery of extensive subterranean glacial ice on Mars verifies that significant amounts of water (in the form of ice) exist below the surface of Mars. But of greater importance, a second paper provides an astounding mechanism for releasing enormous volumes of Flood waters by the violent and instant thaw of glacial ice.

---

311 Keller, C.B. and others. 2018. Neoproterozoic glacial origin of the Great Unconformity. Proceedings of the National Academy of Sciences. 116(4).
312 Nerozzi, S. 2019. Buried Ice and Sand Caps at the North Pole of Mars: Revealing a Record of Climate Change in the Cavi Unit With SHARAD. Geophysical Research Letters. Volume 46, Issue 13.

We suggest that meteoritic impacts on Mars may have repeatedly caused similar liquefaction to enable violent eruption of groundwater. The amount of erupted water may be comparable to that required to produce catastrophic floods and to form outflow channels.[313]...The surface of Mars preserves landforms associated with the largest known water floods... recent spacecraft images document a phase of outburst flooding and associated volcanism that seems no older than tens of millions of years... On both Earth and Mars, abrupt and episodic operations of these megascale processes have been major factors in global climatic change. On relatively short time scales, by their influence on oceanic circulation.... The Martian megafloods are hypothesized to have induced the episodic formation of a northern plains "ocean"...[314]

In a second case, Enceladus is a Snowball moon covered by nearly pure water ice. And as we discussed earlier, jets of water from this moon reach outer space!

At Enceladus, scientists think that hot water containing dissolved silica gushes up into the cooler water of the subsurface ocean. ...And the linear fractures there, now referred to as the "tiger stripes," were as warm as minus 261 Fahrenheit (minus 163 Celsius) in some places.[315] The new observations provide helpful constraints on what could be going on with the underground plumbing -- cracks and fissures through which water from the moon's potentially habitable subsurface ocean is making its way into space.[316]

The Enceladus "tiger stripes" through which the water jets are carried into space, are analogous to linear fractures on Earth and volcanic lineaments. For example, immense fracture zones

---

313 Wang, C. and others 2004. Floods on Mars released from groundwater by impact. Icarus. Volume 175. Pages 551-555.
314 Baker, V. 2009. Megafloods and global paleoenvironmental change on Mars and Earth. Geological Society of America. Special Paper 453.
315 Thompson, J.R. 2017. The Moon with a Plume. NASA. Solarsystem.com
316 Dyches, P. 2016. Enceladus Jets: Surprises in Starlight. Nasa,gov.

encircle the Earth along the ocean floors like the strings on a baseball, as illustrated on the back cover of this book.

Asteroid/Comet Impacts: Tilting of the Planet, Tectonic Mayhem, and Collateral Flood Cycles

Asteroid impact sites have been tied to the Cambrian Explosion of life. Three events, in particular, have been extensively studied including the Acraman Impact,[317] the Shuram event, and the Massive Australian Precambrian/Cambrian Impact Structure (MAPCIS).[318]

> *During the Ediacaran Period, Earth's climatic zonation and controls appear to have undergone a radical change that persisted throughout the Phanerozoic Eon. The change may coincide with the world's greatest negative $\delta^{13}C$ excursion, the Shuram event, here interpreted as the result of a very large marine impact that <u>decreased the obliquity of the ecliptic</u>, causing the Earth's climatic system to adopt its present configuration.[319]*

Evidence supports the triggering of the Cambrian Explosion of life forms by a series of abrupt catastrophic impact events. The MAPCIS is considered speculative at the moment, but it looks like a legitimate prospect.

> *"The impact could have blown a large portion of the ocean into the atmosphere," he says. "This would provide a plausible mechanism for the oxygenation of the oceans."...What's more, several deep canyons were formed*

317 Williams, G.E., and V.A. Gostin. 2005. (Published online 2011). Acraman – Bunyeroo impact event (Ediacaran), South Australia, and environmental consequences: twenty-five years on. Australian Journal of Earth Sciences. Volume 52, 2005, Issue 4-5.
318 Connelly, Daniel P. (2009). The case for a massive Australian Precambrian/Cambrian impact structure (MAPCIS). Geological Society of America Abstracts with Programs. 41 (3): 38.
319 Young, G.M. 2013. Evolution of Earth's climatic system: Evidence from ice ages, isotopes, and impacts. GSA Today. Pages 4-10.

around that time, including in South Australia, California, and Uruguay — all around the time of the Shuram anomaly. "The impact could have temporarily flooded continents and contributed to the formation of these deep canyons, as huge volumes of water invaded the land and returned to the oceans," Young says… According to his theory, the spin axis (or "tilt") of our planet would have shifted drastically upon impact, modifying climate patterns.[320]

Grant Young[321] provides an analysis of the Acraman impact and the impact associated with the Shuram events, as potential triggers of the Cambrian Explosion of life.

The second half of the Ediacaran period began with a large impact - the Acraman impact in South Australia… A few million years later (~570 Ma?) there was a second, deeper and longer-lived world-wide $\delta^{13}C_{carb}$ anomaly (the Shuram anomaly) which coincides with extinction of the acanthomorphic acritarchs. Wide distribution of the Shuram event is exemplified by stratigraphic sections from South Australia, Oman, southern California and South China. The widespread anomaly has been tentatively attributed to a marine impact. During recovery from the Shuram event the enigmatic Ediacaran biota achieved its zenith, only to be extirpated and replaced by a polyphyletic assemblage of shelly animals in what is known as the Cambrian "explosion". This extinction-radiation cycle was preceded by glaciation, another $\delta^{13}C_{carb}$ excursion and the highest $^{87}Sr/^{86}Sr$ values known from marine carbonates. These high Sr ratios have been linked to weathering of extensive tracts of continental crust that were elevated during amalgamation of the supercontinent Gondwana.[322]

---

320 Bontemps, J. 2013. Did a Huge Impact Lead to the Cambrian Explosion? Astrobiology Magazine. Astrobio.net
321 Young, Grant M. 2015. Environmental upheavals of the Ediacaran period and the Cambrian "explosion" of animal life. Geoscience Frontiers. Vol. 6, Issue 4. Pages 523-535.
322 Young, Grant M. 2015. Ibid.

D.H. ALEXANDER

Tectonics drove Cambrian Volcanic Havoc

Peak carbon dioxide levels during the entire span of the Phanerozoic (Cambrian to present) occur in the Cambrian, driven by large releases of upwelling magma from ocean ridges, giving birth to Gondwana.

It has been known for some time that the Cambrian explosion was associated with an increase in temperature as the Earth emerged from an icy era, the Cryogenian (850-635 Ma), informally known as "Snowball Earth". Earth thawed in the Ediacaran era (635-541 Ma), warmed further during the Early Cambrian and then became stiflingly hot during the Dead Interval before cooling again... global warming during the Cambrian was caused by volcanic activity associated with the birth of Gondwana. The new evidence comes from zircon crystals... As well as heating the planet, the extra CO_2 acidified the oceans. Many ocean creatures are sensitive to changes in acidity, and this could help explain the Dead Interval. The volcanism died off once Gondwana had formed, CO_2 levels fell and a huge diversity of reef-based animals appeared.[323]...

The late Neoproterozoic–early Paleozoic was a time of widespread convergent tectonism associated with Gondwanan amalgamation (Cawood and Buchan, 2007). The release of enormous volumes of carbon dioxide at the beginning of the Cambrian was due to global plate tectonics... the Cambrian Period was an interval of extreme greenhouse conditions[324] with the highest modeled

---

323 Gray, B. 2014. Cambrian explosion fuelled by volcanic activity. Geological Society of Glasgow. Geologyglasgow.org.uk

324 McKenzie, R.N. and others. 2014. Plate tectonic influences on Neoproterozoic–early Paleozoic climate and animal evolution. Geology. Volume 42. Pages 127-130.

atmospheric CO_2 concentrations of the Phanerozoic (Berner, 2006)[325]

Did Extraterrestrial Impacts and/or Nuclear Explosions Restart the Earth's Magnetic Field?

Scientific studies of the history of dynamos on other planets and planetesimals just might give insight into the shutdown of the Earth's magnetic field just before the Cambrian Explosion and its restart. And they just might provide other insights into the events leading up to the Cambrian Explosion. As we learned earlier, the Earth's magnetic field almost shut down during the Cambrian, but then it restarted.[326]   And Mars lost its oceans, rivers, and atmosphere because it lost its magnetic field and has never recovered.

Rocky planets like Earth, Mars, Mercury and even the moon get their magnetic fields from the movement of molten iron inside their cores, a process called convection. Packets of molten iron rise, cool and sink within the core, and generate an electric current. The planet's spinning turns that current into a magnetic field in a system known as a dynamo... Once upon a time, Mars had a magnetic field, just like Earth. Four billion years ago, it vanished, taking with it the planet's chances of evolving life as we know it. Now scientists have proposed a new explanation for its disappearance...A model of asteroids striking the red planet suggests that, while no single impact would have short-circuited the dynamo that powered its magnetism, a quick succession of 20 asteroid strikes could have done the job. In the case of Mars *a mega-impact would have flattened out the heat cycle inside the planet, too, snuffing out the dynamo*

---

325 Berner, R.A., 2006, GEOCARBSULF: A combined model for Phanerozoic atmospheric O2 and CO2 : Geochimica et Cosmochimica Acta, v. 70, p. 5653–5664.
326 Bono, R.K. and others. 2019. Young inner core inferred from Ediacaran ultra-low geomagnetic field intensity. Nature Geoscience. Volume 12. Pages 143-147.

within about 20,000 years... Without the cold compress of the mantle to siphon heat away from the core, convection wouldn't have a chance... But left alone, convection would have recovered in the outer parts of the core, and eventually penetrated deep and started the whole core churning again. The Borealis impact would have crippled the dynamo, but not killed it outright. [327] The magnetic field was going haywire during the late Ediacaran by reversing its polarities 20 times faster than it does today. <u>These are signs of imminent geodynamo collapse</u>... The fact that the field got stronger after this time instead of collapsing suggests that nucleation began and gave Earth's dynamo the juice it needed...[328]

Could the Earth's magnetic field reversals be affected by *nuclear chain reactions at the core-mantle boundary*?

Geomagnetic field reversals and changes in intensity are understandable from an energy standpoint as natural consequences of intermittent and/or variable nuclear fission chain reactions deep within the Earth. Moreover, deep-Earth production of helium, having $^3He/^4He$ ratios within the range observed from deep-mantle sources, <u>is demonstrated to be a consequence of nuclear fission.</u> Numerical simulations of a planetary-scale geo-reactor were made by using the SCALE sequence of codes. The results clearly demonstrate that such a geo-reactor (i) would function as a fast-neutron fuel breeder reactor; (ii) could, under appropriate conditions, operate over the entire period of geologic time; and (iii) would function <u>in such a manner as to yield variable and/or intermittent output power.</u>[329]

---

327 Grossman, Lisa. 2011. Multiple Asteroid Strikes May Have Killed Mars's Magnetic Field. Wired.com

328 Ferreira, B. 2019. Earth's Magnetic Field Almost Collapsed 565 Million Years Ago. Vice.com.

329 Hollenbach, D.F. and J.M. Herndon. 2001. Deep-Earth reactor: Nuclear fission, helium, and the geomagnetic field. PNAS. Vol. 20 (8). 11085-11090.

Variable fission chain reactions at the core-mantle boundary could account for geomagnetic field reversals and for cyclical intermittent power; potentially driving the 6 cycles of crustal heating and cooling since the Cambrian. Are core-mantle cycles of nuclear materials buildup, explosions, scatter, and re-accumulation drivers of the 6 cycles of Phanerozoic flooding?

Wobble of the Earth's Axis by Sloshing Water

Could it be that the enormous invasions of the continents by the seas triggered a considerable wobble of the Earth's axis? Recent studies show that the movement of water on the surface of the planet can contribute to the wobble of the Earth's axis.

...the movement of water around the world contributes to Earth's rotational wobbles. Earlier studies have pinpointed many connections between processes on Earth's surface or interior and our planet's wandering ways. For example, Earth's mantle is still readjusting to the loss of ice on North America after the last ice age, and the reduced mass beneath that continent pulls the spin axis toward Canada...[330]...there also exists the distinct possibility that relatively rapid and large amplitude changes in the rotational state could have occurred in association with an "avalanche effect" during which the style of the mantle convective circulation switches from one characterized by significant radial layering of the thermally forced flow, to one of "whole mantle" form (eg. Peltier and Solheim, 1994a,b). This process could conceivably act so as to induce the Inertial Interchange True Polar Wander instability that was suggested initially by Gold (1955) and which has recently been invoked by Kirschvink et al. (1997) as plausibly having occurred in the early Cambrian period of Earth history.[331]

---

330 Adhikan S. and E.R. Ivins. 2016. Climate-driven polar motion: 2003–2015. Science Advances. Volume 2. Number 4.
331 Peltier, W.R. 2007. History of Earth Rotation. In Book: Treatise on Geophysics. Pages 243-293.

The connection between avalanches at the Earth's Core and rapid shifting and movement of vast quantities of ocean waters should not go unnoticed. Could cosmic impacts, the Sauk Transgression and later Flood waters, have led to Core Avalanches? And *could it have been this series of interconnected events that led to the restart of the Earth's magnetic field? Could the wobble of the Earth's axis, the collapse of topographic structures at the core mantle boundary, and the restart of the Earth's magnetic field have been caused by surface waters racing back and forth during a worldwide Flood event? And could accumulation of LLSVPs at the mantle-core boundary under Africa and the Pacific have a significant impact on wobble, the magnetic field, and nuclear explosions?*

Photosynthesis and the Great Oxygen Event

The early Shattered Planet, was dominated by reducing agents such as ferrous iron, hydrogen, and methane. These agents prevented the accumulation of a significant percentage of oxygen in the early atmosphere. But as glaciers covering much of the planet broke down bedrock, their meltwaters carried nutrients to the sea. These nutrients were accompanied by gases adsorbed from the early atmosphere, especially carbon dioxide (CO_2). The CO_2 led to a warming of the environment which accelerated the glacial weathering of the land. The dissolved carbon dioxide reacted with water to form organic sediments and control the pH of the water. You might say that *the garden was being prepared* for planting.

Why then did the Great Oxidation Event take so long to occur? Oxygen had to first overwhelm the flux of reductants (e.g., H_2, CH_4) before it could accumulate in the atmosphere. Photosynthetic O_2 accumulation in the surface oceans may have started off slowly. In the wake of the youngest global glaciation 635 Myr ago (a "Snowball Earth"), a hot climate likely promoted elevated primary productivity and organic matter burial. A significant increase in ocean oxygenation followed shortly after the glaciation. For the first time in Earth's history, the oceans contained enough dissolved O_2 to support large complex animal life. Shortly after the end of *the Snowball glaciation and the increase in ocean oxygenation, the first large complex animals appear in the rock record, including those that could move and prey on other organisms. A series of dizzyingly rapid evolutionary innovations driven by environmental, genetic, and ecological factors then culminated in a "Cambrian Explosion"...[332] The largest genetic study ever performed to learn when land plants and fungi first appeared on the Earth has revealed ...that land plants and fungi evolved much earlier than previously thought — before the Snowball Earth and Cambrian Explosion events — suggesting their presence could have had a profound effect on the climate and the evolution of life on Earth..."[333]*

Oxygen provided by photosynthesis set the stage for the Cambrian Explosion of animal life. Fossils representing many of the 36 existing animal phyla that populate the Earth have been found in early Cambrian rocks. Fossils of other phyla also likely exist in the Cambrian rock record. *Considering the phylogenetic tree implications, the fossil record suggests that all metazoan phyla had originated by the close of the Early Cambrian.[334] Were they planted by a higher intelligence?*

Cambrian Apocalypse: Convergence of Events

Recently, scientists have observed the largest impact crater identified in the Solar System, at greater than 2000 kilometers diameter, in the Aitken Basin at the South Pole of the Moon (image).[335] It is the largest crater discovered to date... larger

---

332 Kendall, Brian. 2013. Earth's Oxygen Revolution. Wat on Earth, uwaterloo.com

333 Hedges, B. 2001. First Land Plants and Fungi Changed Earth's Climate, Paving the Way for Explosive Evolution of Land Animals, New Gene Study Suggests. Science.psu.edu

334 Valentine, J.W. 1995. Why No New Phyla after the Cambrian? Genome and Ecospace Hypotheses Revisited. PALAIOS Volume 10, Number 2.

335 Bartels, M. 2019. Weird 'Anomaly' at the Moon's South Pole May Be a Metal Asteroid's Grave. Space.com.

impact craters may exist hidden on Earth. If the lunar surface is an indicator, Earth has been pummeled far more than thought.

Should evolution's "survival of the fittest" creed be supplemented with an appendage that reads "survival of the catastrophic?" A team of Berkeley researchers, analyzing the history of impact cratering on the moon, has reported a surprising increase in the frequency of impacts over the past 400 million years that may have played a central role in the evolution of life on Earth… The data published by this team show that the impact cratering rate had dropped steadily until the unexpected rise when the impact rate returned to the same levels as 3.5 billion years ago. The sudden increase coincides with the "Cambrian explosion," a period in which life on Earth took off with a dramatic burst in the number and diversity of species.[336]

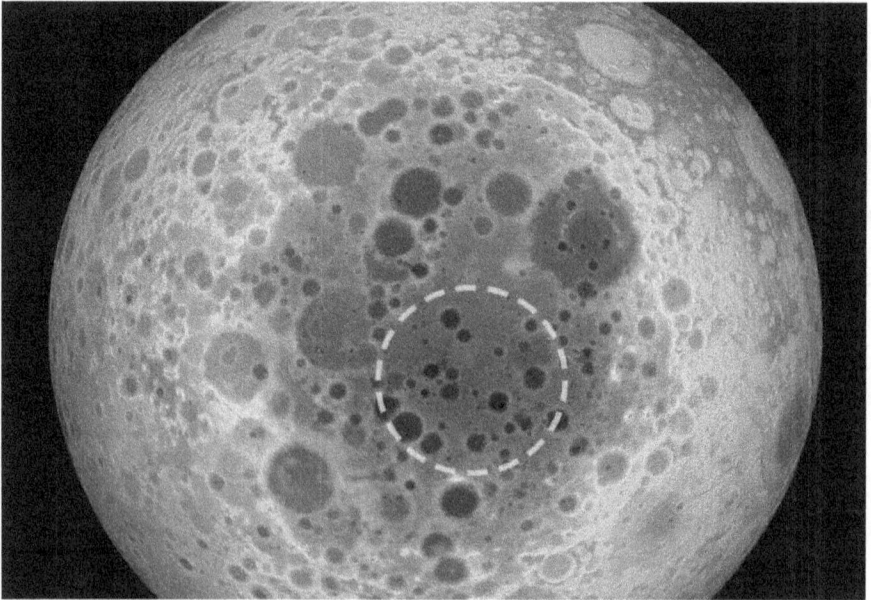

(Image: © NASA/Goddard Space Flight Center/University of Arizona)

---

336 Yarris, Lynn. 2000. Lunar Cratering Shows Surprising Increase During Cambrian Explosion. Lawrence Berkeley Laboratory.

15 *Nuclear Explosions and Impact drivers of Tectonics*

The mountains shall be thrown down, and the steep places shall fall, and every wall shall fall to the ground.[337] The waters thereof roar [and] be troubled, [though] the mountains shake.[338] The earth is broken asunder, the earth is split through, the earth is shaken violently. The earth reels to and fro like a drunkard And it totters like a shack.[339] The channels of the sea appeared, the foundations of the world were discovered.[340] Which removeth the mountains, and they know not: which overturneth them in his anger. Which shaketh the earth out of her place, and the pillars thereof tremble.[341] Thou hast made the earth to tremble; thou hast broken it: heal the breaches thereof; for it shaketh.[342]

What extreme source of energy, forces, and conditions would it take to rapidly mobilize ocean crust and move continents? Knowledge of deep Earth processes are quickly becoming clarified by our satellites. We now have confirmed that the seafloor is born at ocean ridges and consumed at ocean trenches along island arcs and continental coast lines. Our knowledge of the mechanics, physics, and chemistry of the innerworkings of the deep Mantle have become significantly illuminated in this new 21st century, by high pressure-temperature

---

337 Ezekiel 38:20.
338 Psalm 46:3.
339 Isaiah 24:19, 20.
340 2 Samuel 22:16.
341 Job 9:5, 6.
342 Psalms 60:2.

experiments and studies of fluid inclusions. And supporting surface knowledge comes from exploratory drilling and NASA's investigations of the Solar System and the Universe.

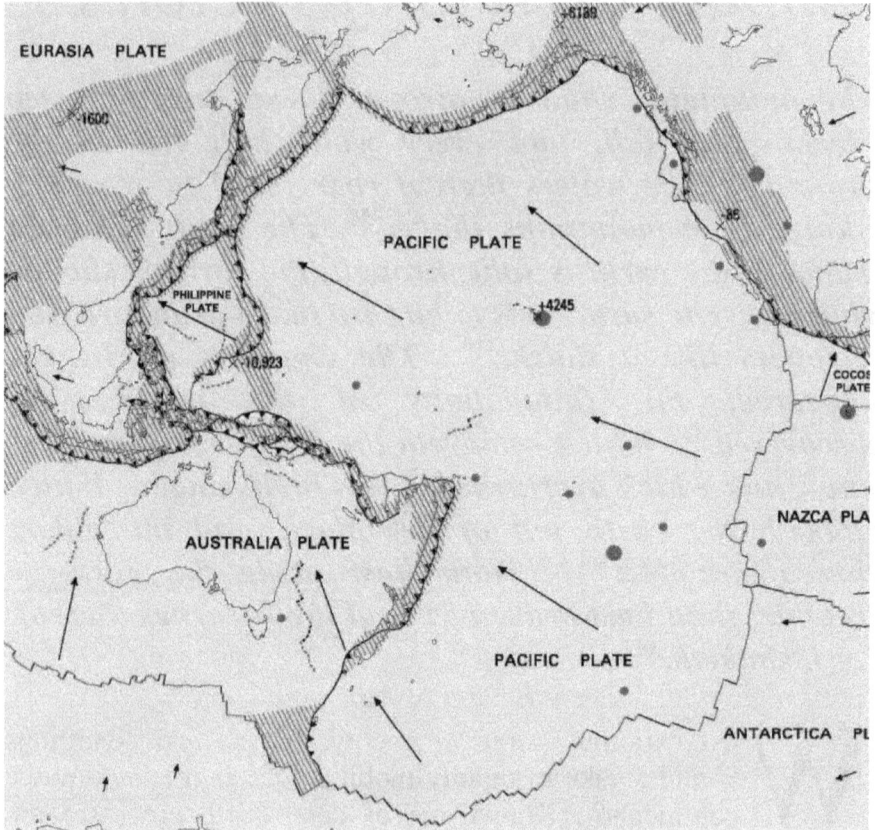

The USGS map[343] illustrates that plate boundaries are defined by earthquakes and volcanism. Importantly, arrows on the map show that the living Planet's tectonic plates are in continuous motion. And as you can see, the shattered plates move; shifting like ice rafts on water. According to the United States Geological Survey:[344]

---

343 Tilling, R.I. et al. USGS. 2006. This dynamic planet: World map of volcanoes, earthquakes, impact craters and plate tectonics. IMAP 2800.
344 Tilling, R.I. et al. USGS. 2006. This dynamic planet: World map of volcanoes, earthquakes, impact craters and plate tectonics. IMAP 2800.

Earth is a dynamic planet, clearly illustrated by its topography, over 1500 volcanoes, 44,000 earthquakes, and 170 impact craters. These features largely reflect the movements of Earth's major tectonic plates and many smaller plates or fragments of plates (including microplates). Volcanic eruptions and earthquakes are awe-inspiring displays of the powerful forces of nature and can be extraordinarily destructive. On average, about 60 of Earth's 550 historically active volcanoes are in eruption each year. In 2004 alone, over 160 earthquakes were magnitude 6.0 or above, some of which caused casualties and substantial damage. Most new crust forms at ocean ridge crests, is carried slowly away by plate movement, and is ultimately recycled deep into the earth--causing earthquakes and volcanism along the boundaries between moving tectonic plates. Oceans are continually opening (e.g., Red Sea, Atlantic) or closing (e.g., Mediterranean).[345]

Clearly the vast amount of Earth's heat comes from radioactive decay. *The vast majority of the heat in Earth's interior—up to 90 percent—is fueled by the decay of radioactive isotopes like Potassium 40, Uranium 238, 235, and Thorium 232 contained within the mantle. These isotopes radiate heat as they shed excess energy and move toward stability. "The amount of heat caused by this radiation is almost the same as the total heat measured emanating from the Earth."[346] Could sudden heat pulses by core-mantle nuclear explosions or asteroid impacts trigger plate tectonics?* According to simulations conducted by O'Neill:[347] *...thermal anomalies produced by large impacts induce mantle upwellings that are capable of driving transient subduction events. Furthermore, we find that moderate-sized impacts can act as subduction triggers by causing localized lithospheric thinning and mantle upwelling, and modulate tectonic activity.* Nuclear explosions would have a similar affect.

---

345 Tilling, R.I. et al. USGS. 2006. Ibid.
346 Anuta, Joe. 2006. Probing Question: What heats the earth's core? psu.edu.
347 O'Neill, C.O., and others. 2017. Ibid.

Slab Pull due to Gravity-Buoyancy a significant force for Plate Tectonics?

Clearly, immense energy triggered from impacts and/or nuclear explosions at the Core-Mantle boundary serve to initiate plate tectonics. But what sustains tectonics? Forsyth and Uyeda[348] found that <u>a slab-pull effect is an order of magnitude stronger than any other force including the push coming from the ridge generating the new ocean floor.</u>

> *A number of possible mechanisms have recently been proposed for driving the motions of the lithospheric plates, such as pushing from mid-ocean ridges, pulling by down-going slabs, suction toward trenches, and coupling of the plates to flow in the mantle. We advance a new observational method of testing these theories of the driving mechanism.* <u>*The results indicate that the forces acting on the down-going slab control the velocity of the oceanic plates and are an order of magnitude stronger than any other force. Namely, all the oceanic plates attached to substantial amounts of down-going slabs move with a 'terminal velocity' at which the gravitational body force pulling the slabs downward is nearly balanced with the resistance acting on the slab;*</u> *regardless of the other features of the trailing horizontal part of the plates.*[349]

Therefore, a reduction in frictional resistance between the new lithospheric slab and the asthenosphere that it rests upon, is key to controlling plate velocity. According to Niu:

> *the above is ascribed to the sinking of the "cold" and "dense" oceanic lithosphere (plates) in subduction zones. In addition, experimental studies and seismology reveal two major phase transitions in the mantle at depths of 410 km*

---

348 Forsyth, D. & Uyeda, S. (1975): On the relative importance of the driving forces of plate motion. – Geophys. J. Int. 43: 163–200.
349 Forsyth, D. & Uyeda, S. (1975). Ibid.

and 660 km (or called 410-D and 660-D seismic velocity discontinuities), between which is called the mantle transition zone, and across which the density of new mineral assemblages become denser with depth. Relative to the 410-D of the ambient mantle, <u>the cold subducting slab changes to a denser mineral assemblage earlier at a shallower depth, gaining excess negative buoyancy, enhancing the Slab Pull,</u> and thus facilitating the plate motion.[350]

Confirmation of a Slippery Zone at the Base of Tectonic Plates and in the Asthenosphere

Are there surfaces above and below tectonic plates that act like décollement surfaces in thrust zones (e.g., analogous to the case of Heart Mountain) that can reduce frictional forces and enable increased acceleration of slab descent velocities? In order to examine the lower crust and asthenosphere boundary, Stern and others[351] applied a unique approach using *reflected seismic waves generated by surface explosives in steel cased boreholes to image the Pacific plate as it descends beneath New Zealand. <u>They found the fabled slippery surface that is essential to plate tectonics.</u> Their seismic results led them to conclude that a narrow, well-defined lithosphere-asthenosphere boundary exists at the base of the crust* that is:

less than 1 kilometre thick at the top of a 10-km-thick channel, in which slow seismic velocities may require the presence of water or melt. The authors suggest that <u>the thin channel decouples the lithosphere from the asthenosphere and allows plate tectonics to take place.</u> The existence of

---

350 Niu, Yaoling. 2014. Geological understanding of plate tectonics: Basic concepts, illustrations, examples and new perspectives. Global Tectonics and Metallogeny 10/1, p. 23–46.
351 Stern, T.A. and others. 2015. A Seismic Reflection Image for the Base of a Tectonic Plate. Nature volume518, pages85–88

such a localized channel probably has implications for the driving forces of plate tectonics and mantle dynamics.[352]

Bercovici[353] proposed another mechanism for lubrication of tectonic plates; a self-lubricating rheological mechanism:

Recent studies suggest that self-lubricating rheological mechanisms are most capable of generating plate-like motion out of fluid flows. The basic paradigm of self-lubrication is nominally derived from the feedback between viscous heating and temperature-dependent viscosity. We propose a new idealized self-lubrication mechanism based on void e.g., pore and/or microcrack generation and volatile e.g., water ingestion.

Transition Zone Water is provided by Sinking Slabs, Serpentinization of Mantle Rocks, and Slab Dehydration Reactions

Importantly, descending slabs gain density or negative buoyancy as they travel downward into the mantle, accompanied by changes in mineral densities: in the Upper Transition Zone at 410 km depth; explicitly olivine (α-Mg_2SiO_4; density 3.27 gm/cc) $\rightarrow$ wadsleyite[354] (β-$Mg2SiO4$; density 3.84 gm/cc) under increasing pressure. This change in mineral chemistry substantially increases the momentum of the sinking slab. As the seafloor slab reaches depths between 520km to 660km another phase transition occurs. At these depths the wadsleyite (orthorhombic) phase of olivine undergoes a transition to ringwoodite which is yet **another even**

352 Rychert, Catherine A. 2015. The Slippery Base if a Tectonic Plate. Nature volume 518, pages39–40.

353 Bercovici, D. (1998): Generation of plate tectonics from lithosphere-mantle flow and void-volatile self-lubrication. Earth Planet. Sci. Lett. 154: 139–151.

354 Houser, C. and Quentin Williams. 2010. Reconciling Pacific 410 and 660 km discontinuity topography, transition zone shear velocity patterns, and mantle phase transitions. Earth and Planetary Science Letters 296(s 3–4):255–266.

higher pressure compositional equivalent of olivine. It incorporates hydroxyl ions into defects in its crystal lattice, leading to the prediction that <u>there is from one to three times the world's ocean equivalents of water in the mantle's Transition Zone</u>.[355]

> *The transition zone slab will undergo isobaric heating with time and release water (most likely in the form of hydrous melt) by changing into progressively water-poor mineral phases. The water (hydrous melt) so released will rise and percolate through the upper mantle to facilitate the formation and maintenance of the Low Velocity Zone [LVZ] which is in fact the process of lithosphere thinning and related magmatism.* [356] *...The high water storage capacity of minerals in Earth's mantle transition zone (410- to 660-kilometer depth) implies the possibility of <u>a deep H_2O reservoir, which could cause dehydration melting of vertically flowing mantle</u>... In experiments, the transition of hydrous ringwoodite to perovskite and (Mg,Fe)O produces intergranular melt. Detections of abrupt decreases in seismic velocity where downwelling mantle is inferred are consistent with partial melt below 660 kilometers... <u>hydration of a large region of the transition zone ... may act to trap H_2O in the transition zone.</u>* [357]

Other mineral transformations are important in transporting nuclear materials to the core-mantle boundary.

> *A phase transition of MgSiO3 perovskite, the most abundant component of the lower mantle, to a higher-pressure form called post-perovskite was recently discovered for pressure and temperature conditions in the vicinity of the Earth's*

---

355 Oskin, Becky. 2014. Rare Diamond confirms that Earth's mantle holds an ocean's worth of water. Scientific American.

356 Niu, Yaoling. 2014. Geological understanding of plate tectonics: Basic concepts, illustrations, examples and new perspectives. Global Tectonics and Metallogeny 10/1, p. 23–46.

357 Schmandt, Brandon; Jacobsen, Steven D.; Becker, Thorsten W.; Liu, Zhenxian; Dueker, Kenneth G. 2014. Dehydration melting at the top of the lower mantle. Science. 344 (6189): 1265–1268.

> *core-mantle boundary. This discovery has profound*
> *implications for the chemical, thermal, and dynamical*
> *structure of the lowermost mantle called the D" region.*[358]

Perovskite incorporates uranium and plutonium within its lattice and is associated with the kimberlites, lamproites, and carbonatites that I studied at McClure Mountain. Could this post-perovskite phase be the key supplier of nuclear material to core-mantle boundary nuclear reactors?

Seismic Tomography shows the Farallon Plate is shoved >1800 miles under North America

Most agree that the subduction of the Farallon Plate beneath western North America during the break-up of Pangea caused the Laramide Orogeny around 40-70 million years ago. As the Farallon Plate dove under the North American Plate it created the block fault and thrust faulted mountains. The subduction of the Farallon plate gave rise to the Rocky Mountains, including the Sangre de Cristo Mountains of Colorado. The event not only resulted in the uplift of mountains but caused volcanism, and associated mineral deposits. The Farallon Plate's presence is felt all the way to the New Madrid seismic zone.[359] How could such an enormous slab of ocean floor maintain its integrity if it were forced through mantle rock?

Impacts, Plate Tectonics, & Mountain Building

Imagine the forces required to form a mountain larger than Mount Everest within ten minutes. According to scientists studying the Chicxulub Crater, *The Asteroid that Killed the Dinosaurs Formed a Mountain Range Taller Than Mount Everest in Less than Ten*

---

358 Hirose, Kei and Thorne Lay. 2008. Discovery of Post-Perovskite and New Views on the Core-Mantle Boundary Region. Elements. Volume 4(3). 183-189.
359 Forte, A.M. and others. 2007. Descent of the ancient Farallon slab drives localized mantle flow below the New Madrid seismic zone. Geophysical Research Letters, Volume 34, Issue 4.

Minutes.[360] Essentially, the enormous asteroid, slammed into the Earth "fluidizing" the impacted solids so that they behaved essentially as a fluid pushing rocks downward and outward.

> *According to geophysicist Sean Gulick: If this deep-rebound model is correct (it's called the dynamic collapse model), then our peak ring rocks should be the rocks that have travelled farthest in the impact - first, outwards by kilometres, then up in the air by over 10km, and back down and outwards by another, say, 10km. So their total travel path is something like 30km, and they do that in under 10 minutes. If you picture all of this happening in a slightly slower-moving fluid than water would be, you can envision that the center that rebounds upwards and splashes upwards would kind of collapse outwards, So just as the sides are falling in, this rebounding center is sort of collapsing outwards to create ... this ring of mountains, made from material that ultimately came from fairly deep.*[361]

Large impacts cause fractured, shocked basement granitic rocks cross-cut by dikes and shear zones, all of which generate large vertical fluxes and increased porosity in the crust.[362] Others conclude that asteroid impacts initiate subduction[363,364] and cause mixing[365] of the mantle. Asteroid impacts provide the energy for catastrophic plate tectonic events linked with the rise of mountains, global flooding, and major extinction events.

---

360 Lee, Rhodi. 2016. Asteroid that Killed the Dinosaurs Formed Mountain Range Taller than Mount Everest in Less than 10 Minutes. Tech Times.
361 Lee, Rhodi. 2016. Ibid.
362 Morgan, J.V. and others. 2016. The formation of peak rings in large impact craters. Science Volume 353, Issue 6314. Pages 878-882.
363 O'Neill, C., 2017. Impact-driven subduction on the Hadean Earth. *Nature Geoscience.* Volume10, pages 793–797.
364 Sleep, N.H. and Donald R. Lowe. 2014. Physics of Crustal Fracturing and Chert Dike Formation triggered by Asteroid Impact, Barberton Greenstone Belt, South Africa. Geochem. Geophys. Geosyst.,15, 1054–1070.
365 Marchi, S. 2014. Widespread mixing and burial of Earth's Hadean crust by asteroid impacts. Nature. Volume 511, pages578–582

Asteroid Impact and/or Nuclear Core-Mantle induced Tectonics: not Uniformitarian Tectonics

Geologists are finding far more evidence of catastrophic impacts; several are much larger than those currently verified. Suspected impact craters in excess of 300 kilometers in diameter include the Arganaty crater of Kazakhystan, the MAPCIS (*Cambrian*; up to 2000 kilometers) structure of Australia, the Bangui Magnetic Anomaly (*Cambrian*) of the Central Republic of Africa, and the Bohemian circular structure of the Czech Republic (late Precambrian to early *Cambrian*) all of which equal or exceed the diameters of the Vredefort (largest verified impact crater), Sudbury (about 250 kilometers), and Chicxulub (about 150 kilometers) impact crater diameters. Multiple lines of evidence point to the rapid breakup of Rodinia followed by the rapid formation of Gondwanaland.

> *Cambrian paleomagnetic data shows anomalously fast rotations and latitudinal drift for all of the major continents. These motions are consistent with an Early to Middle Cambrian inertial interchange true polar wander event, during which Earth's lithosphere and mantle rotated about 90 degrees in response to an unstable distribution of the planet's moment of inertia... Geological evidence points to the breakup of one supercontinent, Rodinia, and the almost simultaneous assembly of another, Gondwanaland...The new ages, along with paleomagnetic data, indicate that continents moved at rapid rates that are difficult to reconcile with our present understanding of mantle dynamics. We propose that rapid continental motions during the Cambrian period were driven by an interchange event in Earth's moment of inertia tensor.[366]*

---

366 Kirschvink, J.L., and others. 1997. Evidence for a Large-Scale Reorganization of Early Cambrian Continental Masses by Inertial Interchange True Polar Wander. Science. Vol. 277, Issue 5325, pp. 541-545.

The rapid breakup of continental plates would result from ginormous seismic waves that would oscillate between the Earth's core and the crust. *Such oscillating waves would produce waves of large domes on the Earth's surface.* But is there a sufficient heat source deep within the Planet that drives the rise of magma in with the cyclic nature of the flooding since the Cambrian? Consider this:

> *Think of the mantle as a planet-sized lava lamp where material picks up heat at the core-mantle boundary, becomes less dense and rises in buoyant plumes to the lower edge of Earth's crust, and then flows along that ceiling until it cools and sinks back toward the core.*[367]

According to the United States Geological Survey, *the East Pacific Rise near Easter Island, in the South Pacific has the fastest rate (more than 15 cm/yr).*[368] <u>Are there alternatives that can result in much faster rates?</u> <u>Impact induced seismicity provides one means of kick starting plate tectonics and causing the global "lava lamp" to rapidly turnover.</u> Are there other potential energy sources that could kick start plate tectonics? *Natural Nuclear Reactors were predicted by Kuroda and one such reactor has been confirmed at Oklo, Gabon.* And a number of scientists are once again proposing that natural nuclear reactors might form at the core-mantle boundary. *Could the recycling of ocean crust provide feed for these reactors? Could these form by a cyclical process of: core-mantle concentration of constituents, generation of heat, explosion and back to concentration of constituents, timed with cyclic periods of flooding since the earliest Cambrian.*

> *...the ejection of terrestrial silicate material triggered by a* <u>*nuclear explosion at Earth's core–mantle boundary (CMB),*</u> <u>*causes a shock wave propagating through the Earth... this*</u> <u>*scenario shows that a shock wave created by rapidly*</u> <u>*expanding plasma resulting from the explosion disrupts and*</u>

---

367 Perkins, S. 2016. A Decades-Long Quest to Drill Into Earth's Mantle May Soon Hit Pay Dirt. Smithsonian.com.
368 USGS. 2014. Understanding Plate Motions. pubs.usgs.gov/gip/dynamic

expels overlying mantle and crust material.[369] *...Nuclear fusion reaction sites on Earth require the following conditions: a large quantity of deuterium (D) atoms in solid state materials, an environment with high temperature and high pressure for overcoming the high Coulomb barrier of the fusion reaction and the presence of a physical catalysis promoting the reactions. Thus Earth's Fe-rich alloy core, with limited U and Th, is a probable site. Here we postulate that the generation of heat is the result of three-body nuclear fusion of deuterons confined in hexagonal FeDx core-centre crystals; the reaction rate is enhanced by the combined attraction effects of high-pressure (~364 GPa) and high-temperature (~5700 K)... The possible heat generation rate can be calculated as 8.12×10^{12} J/m^3, based on the assumption that Earth's primitive heat supply has already been exhausted... Meijer and van Westrenen reported nuclear fission of U and Th as heat generation sources at the mantle boundary within Earth's core, based on the distribution of an isotope of Nd in rocks. Bao noted that there are many heat producing elements (U and Th) in a calcium perovskite reservoir at the base of the mantle.*[370]

Could giant seismic waves be caused by rapidly expanding plasma from a nuclear explosion? Could it be that heat, driven by natural nuclear reactors at the Earth's Core-Mantle behave on a cyclical basis causing the rise of ocean ridges and the outpouring of magmas? Do core temperatures and pressures, when shocked by impact waves, favor nuclear explosions? The LLSVPs would raise temperatures at the Core-Mantle boundary, increasing the likelihood of nuclear scenarios. Could Fukuhara and de Meijer, be right about nuclear transients at the Core-Mantle boundary? Could temperatures and compression due to impact shock waves significantly influence such a cycle?

369 de Meijer, R.J. and others. 2013. Forming the Moon from terrestrial silicate-rich material. Chemical Geology, Volume 345. Pages 40-49.
370 Fukuhara, Mikio. 2016. Possible generation of heat from nuclear fusion in Earth's inner core. Scientific Reports. Volume 6. Article number: 37740.

16 6 *Catastrophic Cycles: Waves of Tectonic Flooding*

Or [who] shut up the sea with doors, when it brake forth, [as if] it had issued out of the womb? When   I made the cloud the garment thereof, and thick darkness a swaddling band for it,   And brake up for it my decreed [place], and set bars and doors,   And said, Hitherto shalt thou come, but no further: and here shall thy proud waves be stayed?[371] ...I [am] the LORD thy God, that divided the sea, whose waves roared[372] ...

Furious forces shook the Earth to its very foundation as the continents moved across the globe.   Land and sea changed places while the once mighty Rodinia broke into fragments. Perhaps the devastation was a blessing in disguise. The dynamo at the heart of the Earth was resuscitated apparently by events associated with the Cambrian tectonic fury, shaking the gut of the planet.   Enormous tectonic ruptures in the Earth's crust broke Rodinia into six primary continental   fragments: Gondwana, Laurentia, Baltica, Siberia, Kazakstania, and China.   These ruptures were created along volcanic rift zones where magma raced upward forcing these continental fragments apart, forming new ocean floors.   And these upwelling magmas released unimaginable quantities of gases of water vapor, carbon dioxide, methane, sulfur, and nitrogen into the oceans and atmosphere. And *these major cycles have been repeated 6 times since the Cambrian Explosion. Was a nuclear explosion at the core of things or was it an impact or a synergistic coupling of both?*

---

371 Job 38:8-11.
372 Isaiah 51:15.

Rodinia →Laurentia →Laurasia →Pangea

During the Paleozoic era, typical among the fragments, Laurentia would reform to become Laurasia and eventually become the western margin of the new supercontinent known as Pangea. Like the other fragments of Rodinia, Laurentia was driven by tectonic forces suffering more than 4 major collisions along its coastlines. Collisions among fragments resulted in the Taconic, Acadian, Alleghanian/Ouachitan mountain building episodes on the Eastern margin of Laurasia (modern day East Coast of the United States from Canada to Georgia) with Gondwana (Northwest Africa and South America) spanning the geologic record from the Ordovician to late Permian. Mountains were readily formed in the thin skinned crust from the powerful forces rising from the Great Deep.

Meanwhile, the Antler event resulted in mountain building from California, through Nevada, to Idaho during the Devonian record of Laurasia. Oceanic crust rammed against a western offshore volcanic island arc (Klamath arc analogous to the modern day Sunda Arc of Indonesia) and the oceanic crust dove beneath (subduction) the island arc off the west coast of Laurasia. The collisions were severe enough, not only to raise mountains but also, to cause mountains to slide away from the collision zone like the Heart Mountain klippe in Wyoming.

*The Hamburg Klippe of the Central Appalachian orogenic belt exposed in eastern Pennsylvania displays a complex record of poly-phase mélange and broken formation development in a **convergent margin setting**. It includes an imbricate stack of tectonic slices, which consist of upper Cambrian to Upper Ordovician deep-water and continental slope sedimentary rocks, emplaced by **gravity sliding** onto the Laurentian passive margin during deposition of the Upper Ordovician Martinsburg Formation.*[373]

---

373 Codegone, G. and others. 2012. Formation of Taconic mélanges and broken formations in the Hamburg Klippe, Central Appalachian Orogenic Belt, Eastern Pennsylvania. Tectonophysics. Volumes 568 -569. Pages 215-229.

6 Catastrophic Cycles of the Phanerozoic

The chart on the following page illustrates a sequence of 6 cycles of flood water transgressions (rising) and regressions (falling) during the Phanerozoic. The Paleozoic portion of the rock record documents 4 cycles of major ocean invasions of Laurasia including the Sauk, Tippecanoe, Kaskaskia, and Absaroka transgressions. The Mesozoic rock record documents 1 additional major ocean invasion of Laurasia known as the Zuni transgression. The 6th cycle of transgression and regression, referred to as the Tejas, is recorded in the Cenozoic rock record.

Carbon Dioxide Degassing and Spreading Rate

The chart documents Phanerozoic levels of carbon dioxide compared to sea level rise and sea level flooding cycles. The shaded CO_2 background is adapted from Royer and others.[374] As noted by researchers attempting to determine atmospheric carbon dioxide levels, the two primary periods of high values appear to be in response to the spreading rate during the breakup of two massive continents, degassing from magmatic activity, and zones of accelerated weathering. Note that the atmospheric carbon dioxide peak values are far, far greater than the current levels of today which are in response to enormous outpouring of volcanic CO_2, especially from creation of new ocean floor as shown on the back cover. Note that the sea levels are similar in shape to the carbon dioxide plot. Carbon dioxide peak waves precede the major rise of flood levels. *Delayed sea level rise is in response to glacial melting in keeping with the warming effect of carbon dioxide. It is hard to imagine that mankind could in anyway raise the carbon dioxide levels to anything close to these past levels associated with continental breakup and formation.*

---

374 Royer, D.L. and others. 2004. CO2 as a primary driver of Phanerozoic climate. GSA Today. Volume 14, Number 3. Pages 4-10. ALSO see Foster, G.L., D.L. Royer, and D.J. Lunt. 2017. Future climate forcing potentially without precedent in the last 420 million years. Nature Communications. Volume 8. Article Number 14845.

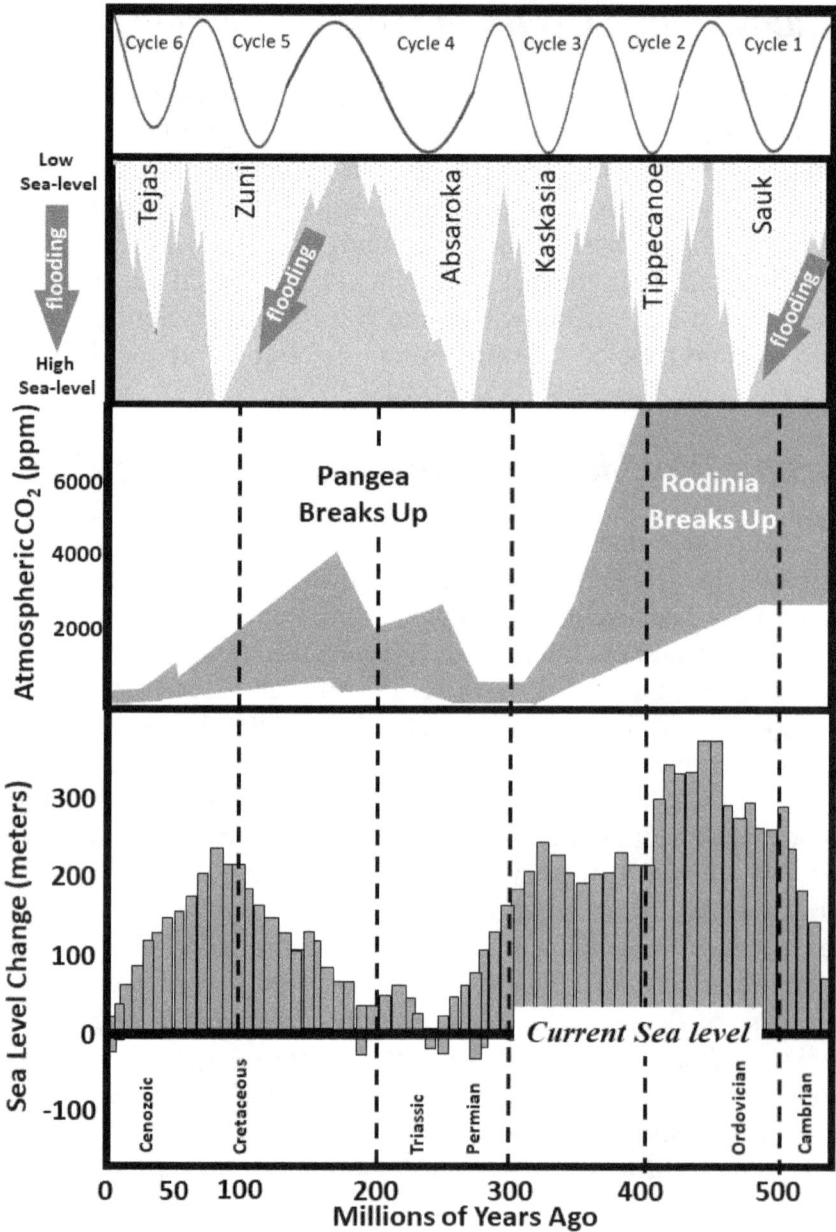

The two primary periods of high carbon dioxide mirror sea level rise and appear to be associated with the break-up of Rodinia and Pangea.

The constant proportionality of weathering and hydrothermal exchange through these geological changes implies that a global feedback exists between weathering and seafloor spreading rates... Such feedback was predicted by the "spreading rate hypothesis," in which any increase in seafloor spreading rate is accompanied by a higher flux of CO_2 degassing from magmatic activity in spreading centers and subduction zones, leading to global warming, higher water acidity, and a global increase in weathering rates. A second possible feedback mechanism is that higher seafloor spreading rates will be accompanied by faster plate convergence, leading to the buildup of relief and faster erosion and weathering rates. [375]

Oxygen Isotope Data: Mid-Ocean Ridge Focus of Water-Rock interactions since Cambrian

A number of recent scientific investigations have focused on seawater temperatures as a critical factor for constraining sea life since the Early Cambrian. The critical threshold of 38 °C stands as the thermal limit for life in the sea. Importantly, they conclude that seawater temperatures have been warm with a constant seawater $\delta^{18}O$ throughout the last ~450 million years. [376]

Our new data can therefore be used to infer a minimal depletion in early Cambrian $\delta^{18}O_{seawater}$ relative to today of about -3‰. With this presumption, our most pristine $\delta^{18}O_{phosphate}$ values translate into sea surface temperatures of about 30 °C indicating habitable temperatures for subequatorial oceans during the Cambrian Explosion. [377]

---

375 Ryb, U. and J.M. Eiler. 2018. Oxygen isotope composition of the Phanerozoic ocean and a possible solution to the dolomite problem. PNAS. Volume 115. Number 26. Pages 6602-6607.

376 Grossman, E.L. and others. 2015. American Geophysical Union, Fall Meeting 2015, abstract id. PP31E-04.

377 Wotte, T. and others 2019. Isotopic evidence for temperate oceans during the Cambrian Explosion. Scientific Reports. Volume 9. Published online.

Benthic metazoans were able to thrive at temperatures of 35-40 °C during intervals of the early and possibly the latest Paleozoic when CO_2 levels were likely 5-10× higher than present-day values. Equally important, there is no resolvable trend in seawater oxygen isotope ratios ($\delta^{18} O$) over the past ~500 million years, indicating that the average temperature of oxygen exchange between seawater and the oceanic crust has been high (~270 °C) since at least the early Paleozoic, which points to <u>mid-ocean ridges as the dominant locus of water-rock interaction over the past half-billion years</u>.[378]

Scientists more recently are pointing to an entire suite of associated events to provide the basis for the 5 major periods of mass extinctions since the Cambrian, pointing to volcanism and tectonism as the primary interconnected factors.

Most extinctions are associated with global warming and proximal killers such as marine anoxia (including the Early/Middle Cambrian, the Late Ordovician, the intra-Silurian, intra-Devonian, end-Permian, and Early Jurassic crises). Many, but not all of these are accompanied by large negative carbon isotope excursions, supporting a volcanogenic origin. [379] *The consistent association of large magmatic provinces (large igneous provinces and continental flood-basalt provinces) with all but one (end-Ordovician) of the five major Phanerozoic mass extinctions suggests that volcanism played a major role. Faunal and geochemical evidence from the end-Permian, end Devonian, end-Cretaceous and Triassic/Jurassic transition suggests that the biotic stress was due to a lethal combination of tectonically induced hydrothermal and volcanic processes,*

378 Henkes, G.A. and others. 2018. Temperature evolution and the oxygen isotope composition of Phanerozoic oceans from carbonate clumped isotope thermometry. Earth and Planetary Science Letters, Volume 490, p. 40-50.
379 Bond, D.P.G. and S.E. Grasby. 2017. On the Causes of Mass Extinctions. Palaeogeography, Palaeoclimatology, Palaeoecology. Volume 478. Pages 3-29.

leading to eutrophication in the oceans, global warming, sea-level transgression and ocean anoxia.[380]

In my opinion, there is a case to be made that the majority of these extinctions are a continuum, especially of sea life, during the repeated invasion of the oceans over 6 primary cycles including the Sauk, Tippecanoe, Kaskaskia, Absaroka, Zuni, and the Tejas transgressions. This continuum is punctuated by peaks of mass extinctions connected to catastrophic events including asteroid impacts and or nuclear explosions, plate tectonics, and volcanism.

A correlation between global marine regressions and mass extinctions has been recognized since the last century and received explicit formulation, in a model involving habitat-area restriction, by Newell in the 1960s. Since that time attempts to apply the species-area relation to the subject have proved somewhat controversial and promoters of other extinction models have called the generality of the regression-extinction relation into question. Here, a strong relation is shown to exist between times of global or regional sea-level change inferred from stratigraphic analysis, and times of high turnover of Phanerozoic marine invertebrates, involving both extinction and radiation; this is valid on a small and large scale. In many cases the most significant factor promoting extinction was apparently not regression but spreads of anoxic bottom water associated with the subsequent transgression.[381]

Geologic evidence from strata in northeastern Siberia demonstrates that the Earth was undergoing life-threatening trauma during the Cambrian. Extremely rapid geomagnetic reversals, even in the

---

380 Keller, G. 2005. Impacts, volcanism and mass extinction: random coincidence or cause and effect? Australian Journal of Earth Sciences. Volume 52. Pages 725-757.

381 Hallum, A. 1989. The case for sea-level change as a dominant causal factor in mass extinction of marine invertebrates. Phil. Trans. Royal Society of London. 325. Pages 437-455.

context of geologic time, are consistent with a hyperactive geodynamo undergoing rapid reversals.

> *We present new magnetostratigraphic results obtained for the Drumian stage (Middle Cambrian) from the Khorbusuonka sedimentary section in northeastern Siberia... The directions from 437 samples define a sequence of 78 magnetic polarity intervals, 22 of which are observed in a single sample. This is an extreme reversal rate, similar to that reported for the Late Ediacaran (late Precambrian)... and proposed to be potentially linked to a late nucleation of the inner core. The reversal frequency appears to have drastically dropped... during... the Furongian/Upper Cambrian.[382]*

Former researchers of the **Cambrian Record** in Siberia corroborate the recent results of Gallet and others referenced above.

> *While studying these magnetic records in 550-million-year-old, ... sedimentary rocks in the Ural Mountains in western Russia, the team discovered evidence to suggest the reversal rate then was 20 times faster than it is today.[383]*

Scientists use magnetometers towed behind ships to map the paleomagnetic field direction recorded in the seafloor rock. The zebra pattern of magnetic stripes are recorded and compiled like a strip chart recorder, capturing the rapidly reversing direction of the Earth's magnetic field as flow upon flow is extruded. Could it be that these individual flows were produced by rapid reversals of the Earth's magnetic field as it underwent *fibrillation*? Were the pulses in response to the deep seated seismic waves oscillating through the Earth? Did such a period of rapid *fibrillation* occur

---

382 Gallet, Y. and others. 2019. Extreme geomagnetic reversal frequency during the Middle Cambrian as revealed by the magnetostratigraphy of the Khorbusuonka section (northeastern Siberia). Earth and Planetary Science Letters. Volume 528.
383 Randall, I. 2016. Hyperactive magnetic field may have led to one of Earth's major extinctions. Science Magazine.org

during the time when asteroid impacts induced seismicity, causing the Earth to reel and quake with massive tremors? *Could the entirety of the Cambrian Apocalypse have taken place as a continuum of tectonically induced waves? In one context, the Cambrian Apocalypse could be considered a miracle. WHY? If the Cambrian Apocalypse had not occurred, it is probable that life on Planet Earth would have died and gone the way of Mars without the restart of Earth's magnetic field.*

Cambrian Apocalypse as a Phanerozoic Model for Waves of Tectonically induced Flooding

The dynamic interplay of the primary events of the *Cambrian Apocalypse presented in the preceding pages reveals an underlying pattern of catastrophism* that is repeated in distinct cycles involving extraterrestrial impacts, plate tectonics, volcanic activity, the flooding of the continents by the seas, and mass extinctions. And perhaps these cycles are tied to *periodic shock waves created by rapidly expanding plasma resulting from nuclear explosions.*

We have discussed the six global cycles of apocalyptic proportion. And by projection there is a short cycle that involves Buttercup and a 7th Cycle that involves you and I. But before we discuss the 7th Cycle we provide evidence to confirm the pattern established by the Cambrian Apocalypse.

17 *6ᵗʰ Cycle: Apocalyptic Pattern Confirmed*

...the mountains shall be molten ... and the valleys shall be cleft, as wax before the fire, [and] as the waters [that are] poured down a steep place.[384] [for it is] the LORD... Who layeth the beams of his chambers in the waters:[385] [Yet] we know that the whole creation groaneth and travaileth in pain together until now.[386] So shall it be at the end of the world[387] [Is] not my word like as a fire? saith the LORD; and like a hammer [that] breaketh the rock in pieces?[388] Which removeth the mountains, and they know not: which overturneth them in his anger. Which shaketh the earth out of her place, and the pillars thereof tremble.[389]

L ight, like a rapidly expanding ball of fire, was accompanied by shock waves as it barreled across the horizon, as if the Sun was descending from its throne where it had reigned from Creation upon high. But the asteroid was not alone. Planet Earth, the center of life in the Universe, was once again under attack. The cosmic bombardment assaulted Earth in waves throughout the 6ᵗʰ Cycle we call the Cenozoic. An incoming giant asteroid hurtled towards Mexico's Yucatan Peninsula, followed later by waves of asteroids that smashed into the Laurentide Ice Sheet, releasing enormous floods. These were followed by even later impacts on Greenland, South Africa, and by all accounts, Siberia. Exploding meteorites disintegrated in air bursts as they

---

384 Micah 1:4.
385 Psalm 104:1-3.
386 Romans 8:22.
387 Matthew 13:49.
388 Jeremiah 23:29.
389 Job 9:5, 6.

sped towards the planet at supersonic speeds, sending fragments, faster than shrapnel from a landmine, flying in all directions. Needless to say, herds of animals raced for safety…but no refuge could be found, even in the bulwarks of the mountains. For the mountains trembled like reeds in the wind and the rocks of mighty caves crumbled like dust. Fires erupted everywhere as black ash swirled in the air and ascended into dark clouds as the once blinding, brilliant-white light, rapidly faded into darkness. Forests were uprooted and floated as giant rafts before being buried beneath the muck that inundated the continents. Majestic forests were laid flat like at Tunguska, buried, and metamorphosed to coal. Forests are even now found frozen beneath the Antarctic. Pounding rains washed layer upon layer of rising dust onto the land surface, even depositing an iridium-rich clay layer around the entire globe, as a fingerprint of the catastrophe in the rock record.

Within the Earth's belly, seismic waves raced radially outward in all directions like ripples upon the surface of a pond. At the impact site, ***it was as if a tsunami sped outward liquefying rock*** as it exploded across the land's surface. The shock waves traversed hundreds of kilometers into the mantle, as they raced around the planet breaking the Earth's surface from below, heaving the once solid land surface into waves like an angry sea. Hundreds of kilometers below the Earth's surface, seismic waves crisscrossed the transition zone between the upper and lower mantle shattering, fracturing, and weakening Earth's crust, opening paths for shallow magma chambers and deep mantle Superplumes to release their ***effervescing molten magma*** to the Earth's surface, oceans, and atmosphere. All around Planet Earth, vibration of these agitated chambers caused magma and gases to burst forth on the surface along shattered seams. Simultaneously, half way around the world, chambers beneath the Deccan Plateau and the ocean ridges surged in response, sending pulse upon pulse of basaltic magma ***as wax before the fire, [and] as the waters [that are] poured down a steep place,*** cascading across the ocean floor as if throbbing to the rhythmic beat of an irreverent drummer. Fractures and faults were moved angrily along mid-ocean ridges like the reopening of an old familiar wound, spurting out jets of bloody red magma and steam;

whilst tsunamis, driven by the endless pulse of massive earthquakes, relentlessly scoured the Earth's surface.

> *...observations suggest the following sequence of events at the end of the Cretaceous period. Just over 66 million years ago, the Deccan Traps start erupting – likely initiated by a plume of hot rock rising from the Earth's core, similar in some ways to what's happening beneath Hawaii or Yellowstone today, that impinged on the side of India's tectonic plate. The mid-ocean ridges and dinosaurs continue their normal activity... years later, Chicxulub hits off the coast of what will become Mexico. The impact causes a massive disruption to the Earth's climate, injecting particles into the atmosphere that will eventually settle into <u>a layer of clay found across the planet</u>... we estimate that around the time of the Chicxulub impact, on the order of 23,000 to 230,000 cubic miles of magma erupted out of the mid-ocean ridges... This is on par with the largest eruptive events in Earth's ... history, including the Deccan Traps... In the end, three-quarters of the Earth's plant and animal species have disappeared; the only remaining dinosaurs are the feathered, flying variety, normally referred to as birds.*[390]

The 6th Cycle, Impacts, Volcanics, Antipodal Flood Basalts, Tejas Flooding, and Glaciation

The Cambrian Apocalypse set the pattern for all the successive cycles. The story of the 6th cycle begins at a famous juncture in time, known as the K-T boundary (Cretaceous-Tertiary), the point in time made famous by the Chicxulub impact story. The pattern of an asteroid impact setting off mass extinctions and the invasion and retreat of seas upon continents is repeated for the sixth time since the Cambrian explosion.

---

390 Karlstrom, L. and J. Byrnes. 2018. The Meteorite That Killed the Dinosaurs May Have Also Triggered Underwater Volcanoes. SMITHSONIAN.COM.

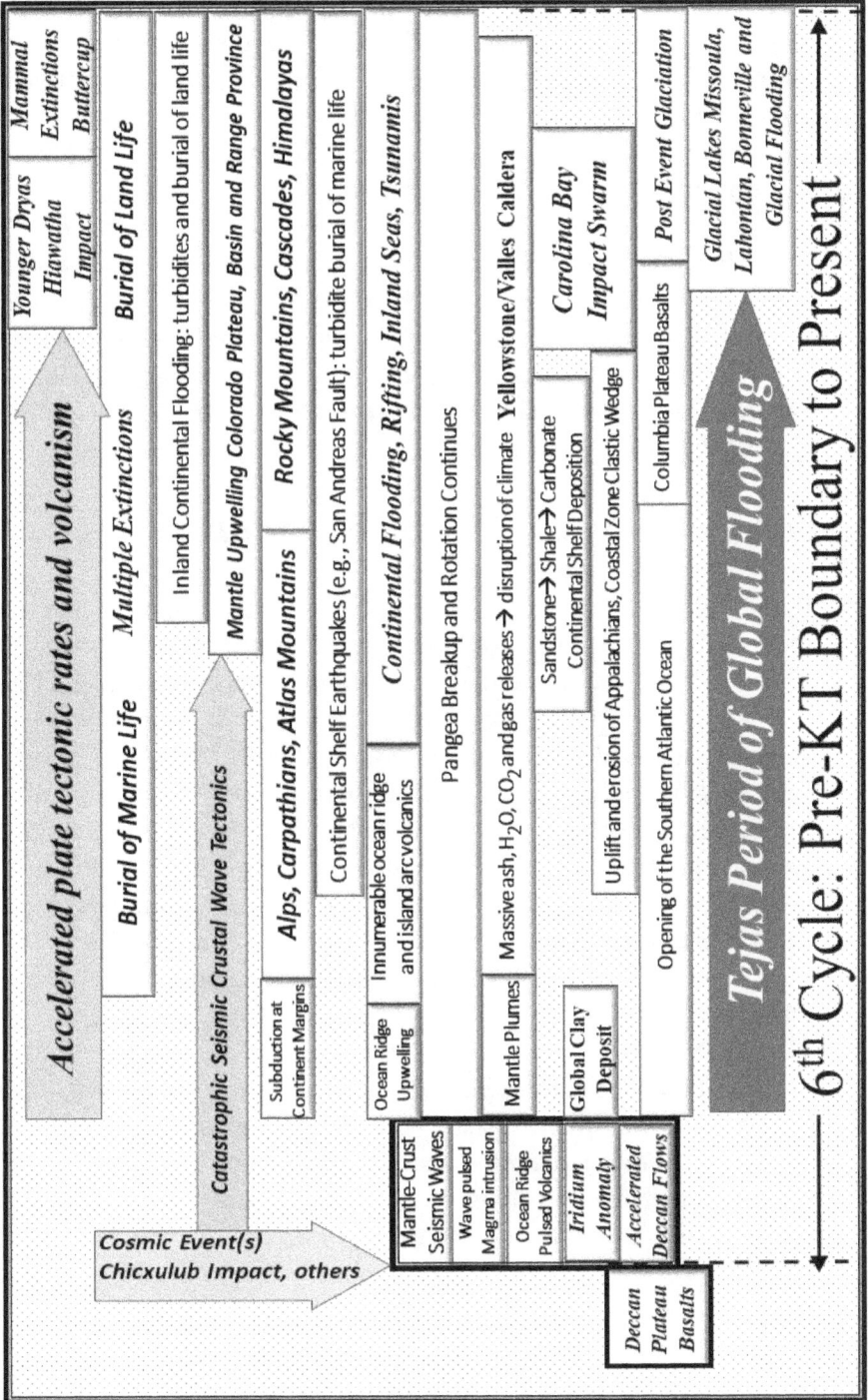

The pattern we constructed for the Cambrian Apocalypse is eerily repeated. Repetitions of flooding cycles appear to be a "continuum" of associated events. Coincidentally, scientists have concluded that asteroid impacts substantially increased from the Cambrian Explosion of Life to the Present day establishing a rhythm for the successive patterns.

> *Mazrouei et al. …found that the impact rate increased within the past ~500 million years, a conclusion strengthened by an analysis of known impact craters on Earth. Crater size distributions are the same on Earth and the Moon over this period, implying that terrestrial erosion affects all craters equally, regardless of their size.*[391]

Asteroid Impact: Seismically Induced Tectonics

Large asteroid impacts can cause a crust busting event that mechanically sets forth seismic waves rippling through the Planet. These waves ricochet off the core and lift the crust from beneath like an explosion below a sheet of ice, fracturing the crust and shoving the fragments in lateral motion. Vertical motion is triggered as these seismic waves intersect subsurface chambers of pressurized molten liquid and gas, violently shaking them and freeing plumes, even Superplumes along the weakened zones among plates, like pulling the cork from a champagne bottle. These magmas, supercharged by supercritical fluids, buoyantly race upward between continent-sized fragments setting plate tectonics in motion.

> *Eruptive phenomena at all scales, from hydrothermal geysers to flood basalts, can potentially be initiated or modulated by external mechanical perturbations. We present evidence for the triggering of magmatism on a global scale by the Chicxulub meteorite impact at the Cretaceous-Paleogene (K-Pg) boundary, recorded by transiently*

---

391 Mazrouei, S. and others. 2019. Earth and Moon impact flux increased at the end of the Paleozoic. Science. Volume 363, Issue 6424, Pages 253-257.

increased crustal production at mid-ocean ridges. Concentrated positive free-air gravity and coincident seafloor topographic anomalies, associated with seafloor created at fast-spreading rates, suggest volumes of excess magmatism in the range of ~10⁵ to 10⁶ km³ . Widespread <u>mobilization of existing mantle melt</u> by <u>post-impact seismic radiation can explain the volume and distribution of the anomalous crust.</u> This massive but short-lived pulse of marine magmatism should be considered alongside the Chicxulub impact and Deccan Traps as a contributor to geochemical anomalies and environmental changes at K-Pg time.[392]

Antipodal Catastrophic Tectonics and Volcanism

Imagine a sphere for a moment, perhaps an orange. Looking at the "pole" of the orange where it was once attached to a stem, draw imaginary lines from this pole. You'll soon recognize that all these lines intersect at the opposite pole of the orange. The first point is opposite it's antipode. Now imagine seismic waves racing around Planet Earth in all directions, from a point like the lines of longitude on a globe, triggering thousands of earthquakes like the 2010 event that destroyed Haiti. Assuming, for purposes of simplification, that all waves travel at similar velocities. If so, they will all converge at the antipode over a narrow range of time! Evidence of this has been observed on Mars:

> *The regions antipodal to Mars' three largest impact basins, Hellas, Isidis, and Argyre, were assessed for evidence of impact-induced disrupted terrains…The results suggest that if antipodal fracturing were associated with later volcanism, then Alba Patera may be related to the Hellas event, as proposed by Peterson (Lunar Planet. Sci. 9, 885-886, 1978)). <u>Alba Patera is a unique volcano in the solar system, being a shield volcano which emitted large volume lava flows.</u> This*

---

392 Karlstrom, L. and J. Byrnes. 2018. Science Advances. Volume 4. eaao2994.

volcanism could be the result of the focusing of seismic energy which created a fractured region that served as a volcanic conduit for the future release of large volumes of magma.[393]

Undoubtedly, numerous large impacts have caused antipodal tectonic events on Earth like the synergism that connected the Chicxulub impact with the accelerated release of the Deccan Plateau Flood Basalts half way around the world. The convergence of waves at the antipode initiated fracture zones on the opposite side of the globe facilitating fracture and fault pathways for the outpouring of the enormous Deccan trap flood basalt flows[394] as illustrated on the front cover.

... using the Chicxulub impact as our case study. Numerical simulations are based on a spectral-element method, representing the impact as a Gaussian force in time and space. Simulating the impact as a point source at the surface of a spherically symmetric earth model results in deceptively large peak displacements at the antipode... based on a point source in a spherically symmetric earth model, wave interference enroute to the antipode induces 'channels' of peak stress that are five times greater than in surrounding areas. Underneath the antipode, we observed 'chimneys' of peak stress, strain and velocity, with peak values...[395]

According to Karlstrom and Byrnes, the seismic energy greatly affected the outpouring of enormous volumes of magma: *widespread mobilization of existing mantle melt by post-impact seismic radiation can explain the volume and distribution of the anomalous crust.*

393 Williams, D.A. and R. Greeley. 1994. Assessment of Antipodal-Impact Terrains on Mars. Icarus. Volume 110, Issue 2. Pages 196-202.
394 Sanders, R. 2015. Asteroid impact, volcanism were one-two punch for dinosaurs. News.berkeley.edu.
395 Meschede, M.A. and others. 2011. Antipodal focusing of seismic waves due to large meteorite impacts on Earth. Geophysical Journal International. Volume 187. Issue 1. Pages 529-537.

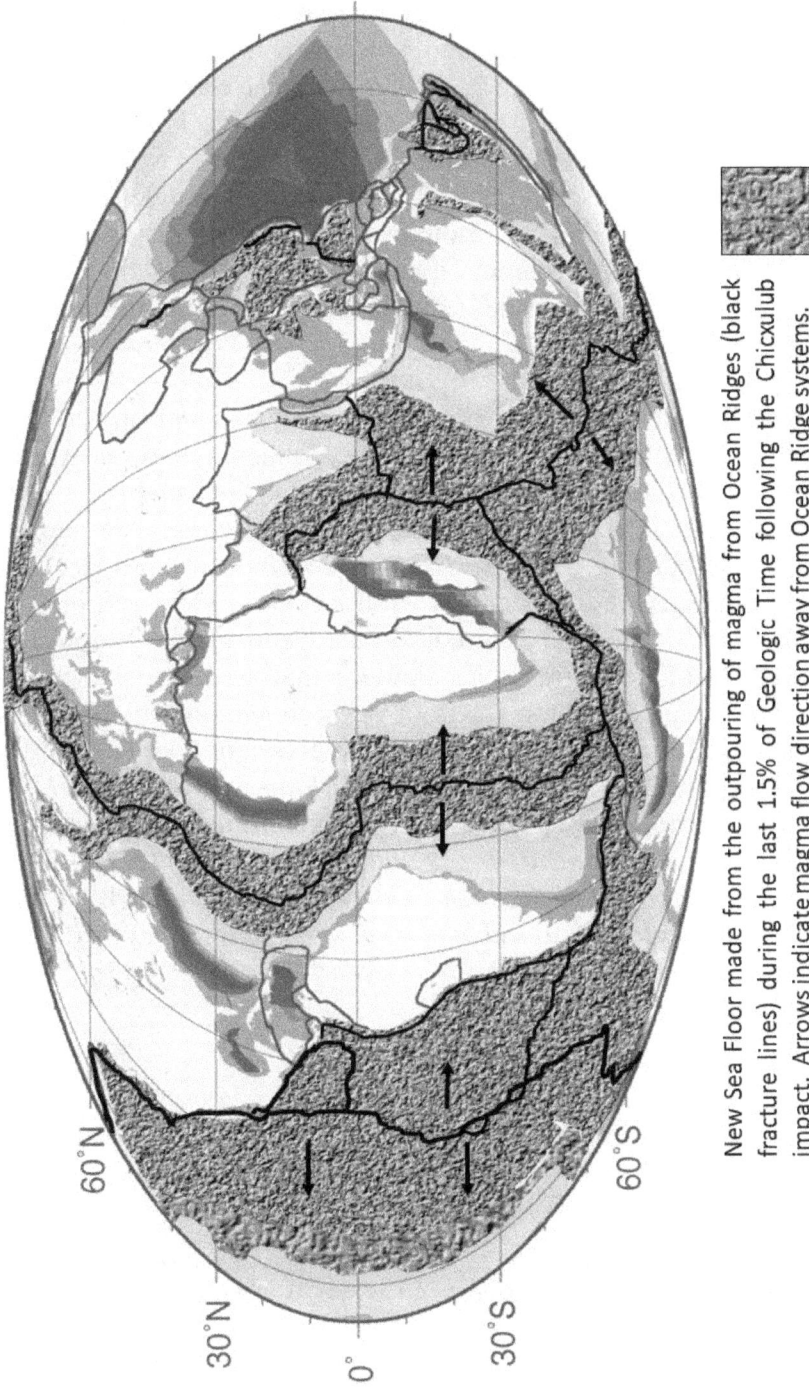

New Sea Floor made from the outpouring of magma from Ocean Ridges (black fracture lines) during the last 1.5% of Geologic Time following the Chicxulub impact. Arrows indicate magma flow direction away from Ocean Ridge systems.

Based on their research they suggest that the Chicxulub impact, instigated worldwide rifting and volcanism in ocean basins as a collateral effect. And there is further evidence that the Chicxulub impact may not have been alone. Some scientists believe that the dinosaurs were killed by a swarm of comets. Two additional contemporary impact craters are found in the Ukraine[396] and the Gulf of Mexico[397] that are thought to be part of the swarm. A swarm of asteroids or comets over a short period of time would significantly compound the seismic energy racing through the planet.

The map of the age of the ocean floor[398] (here and on the back cover) illustrates the enormous quantity of magma that was erupted to the surface of the ocean floor during the relatively short geologic time interval referred to as the Cenozoic. Perhaps as much as a half of the seafloor was continuously-produced by nearly uninterrupted volcanism along ocean ridges during the Cenozoic rock record.[399] Incredibly, the Cenozoic only represents about 1.5% of geologic time. And of that 1.5% "time" period, more was erupted as new sea floor at the beginning of the period than later in the period, suggesting an injection of energy followed by the decay of that energy. *This fits with an asteroid impact that sets off seismic waves that decay with time as the mantle temperatures drop and a steady state is restored. This also fits with a nuclear explosion at the Core-Mantle boundary. Could both an impact and a nuclear explosion have occurred as a single coupled event?*

396 Tytell, D. 2002. Did a Comet Swarm Kill the Dinosaurs? skyandtelescope.com
397 Keller, G. and others. 2002. Multiple impacts across the Cretaceous– Tertiary boundary. Earth-Science Reviews. Volume 62. Pages 327 – 363.
398 Muller, R.D. and others 2008. Age, spreading rates and symmetry of the world's ocean crust. Geochem. Geophys. Geosyst,, Volume 9. QO4006. As adapted by NOAA, (NGDC).
399 Xu, X. and others. 2006. Global reconstructions of Cenozoic seafloor ages: Implications for bathymetry and sea level. Earth and Planetary Science Letters Volume 243. Pages 552–564

COMING EVENTS OF THE 7TH APOCALYPSE

And the stars of heaven fell unto the earth... and every mountain and island were moved out of their places. And the kings of the earth... and every free man, hid themselves in the dens and in the rocks of the mountains...[400] *the first heaven and the first earth were passed away; and there was no more sea...*[401]*the heavens shall pass away with a great noise, and the elements shall melt with fervent heat, the earth also and the works that are therein shall be burned up.*[402] *But of that day and [that] hour knoweth no man...*[403]

With our Apocalyptic Pattern confirmed, we can view the events of the current day in the context of the 7th Apocalypse: ***APOCALYPSE NOW.*** The **Tejas flood sequence** was the last period of flooding across North America; ending about 24 million years ago, during a period of increasing glaciation. We are now in a period of warming, glacial melting, and rising seawaters. What will the future bring? Is a catastrophic impact or nuclear event about to usher in a 7th Apocalypse as the rock record and prophecy both predict?

---

400 Revelation 6:13-15.
401 Revelation 21:1; 2 Peter 3:5-7.
402 2 Peter 3:9-12.
403 Mark 13:32.

18   1st *Extinction of Man and the Demise of Buttercup*

...it shall come as a destruction...Therefore shall all hands be faint, and every man's heart shall melt: And they shall be afraid: pangs and sorrows shall take hold of them;[404] *And the serpent cast out of his mouth water as a flood after the woman, that he might cause her to be carried away of the flood*[405]*... and as it were a great mountain burning with fire was cast into the sea: and the third part of the sea became blood; and the third part of the creatures which were in the sea, and had life, died...And the name of the star is called Wormwood: and the third part of the waters became wormwood; and many men died of the waters, because they were made bitter.*[406]

The hunting season on the high plains was at its peak and had already been deemed a huge success. It would only be a few more days before they broke camp and headed to their winter homes. Hunting parties were returning to camp, dragging litters with their bounty of hides, tusks, and meat for their community. By nightfall, the skins of the prized mammoth were scraped, cleaned, and hung to cure. Within hours, the meat was stored in a smokehouse to dry. The fires were lit and the community would celebrate the hunt and praise the god's for their success. During their journey, the men of the hunting party had watched warily as a glowing *"snake"* god moved through the heavens. They were somewhat unnerved because the scene in the heavens even spooked the herds.

---

404 Isaiah 13:6-8.
405 Revelation 12:15.
406 Revelation 8:8-11.

Meanwhile, back at the camp, the "*shaman*" had also been making his observations. He interpreted that this new "*snake*" god had great powers, since its great light caused the light of the Scorpion, the Wolf, and the Vulture gods to grow dim. The hunters approached the shaman, the village wiseman, to understand the heavenly scene that they had watched in nights past. The shaman told them that he also was troubled by what he saw, saying that it was a sign that the Earth itself was in great danger! Yet he knew they were powerless against this god, so he warned them to bow down and make preparations to welcome the snake god's arrival. But the men scoffed at him. Afterall, their gods had rewarded them with a great bounty. They prepared for the celebration of their great hunt but each one harbored fears within his heart. And the shaman continued to wonder at the unrivaled power of this great light in the heavens. What could it mean? Would it cause destruction? What would the future hold for the tribe?

Celebrations were planned and offerings were being readied so that this new powerful god would be appeased. The tribe had high hopes that the snake god would give them powers that would make them superior to their rivals. Some of the men put on crudely crafted head dresses with antlers while another wore that of a snake with a great long tail. As they danced, they choreographed their version of the ongoing war in the heavens. Over the days that followed, they watched as the war in the heavens continued to escalate. They became accustomed to its presence and accepted it as a new god to worship.

The Last Hunt

In the morning, the hunters headed back on their trek of several days to the forests and plains where the herds roamed. In the distance, Buttercup watched as strange figures appeared along the tall grasses and tree line that bordered the tundra meadow. These strange figures were covered with fur and antlers and their camouflage didn't alert the herds. Their skins smelled like the other members of the herds and consequently, they didn't alert the suspicion of the herd bulls that served as sentries. Perhaps the greatest predator of the herds weren't the sabre-tooth tigers or the vicious wolf packs; but these strange stealthy upright figures that moved through the shadows. They were adept at stalking the great herds of wooly mammoths, horses, muskox, and reindeer across the alpine savanna covered with wildflowers, grasses shrubbery, sedges and flowers like buttercups, poppies, and anemones that covered the ground. These strange upright creatures wielded clubs and spears tipped with stone.

It was a peaceful day with a gentle breeze as the watchers circled the herd. All the while, the herds were unaware of the impending danger. But not even the stalkers were aware of the greater danger that was rapidly approaching in the sky; they were focused on the hunt. Food was plentiful and the herds were putting on fat for the coming winter. Rays of sunshine streaked across the meadows filtered by the clouds, adding to the serenity of the environment. But everything was about to change. ***The snake god began casting fire and brimstone upon the Earth.***

Instantly without warning, flashes of blinding light streaked across the sky as though the sun had exploded. The streaks of light were followed by loud sonic booms like the 2013 bolide explosion above Chelyabinsk. The light, the sound, and the shaking ground were followed by loud animal screams of stampeding herds. Tiny fragments of raining fire struck animals within the herds[407] adding to the frantic flight for life. Towering clouds of ash jetted high into the sky and the daylight suddenly became eerily dark as the ash blocked out the sun. The smell of sulfur and other gases gave off the distinctive stench of burnt flesh. Moments later, gale force winds raised havoc within the herds, separating the young from the adults. The herds sensed the rumbling sound and ran for high ground as the earth beneath their hooves vibrated and groaned. But run as they may, it was no use. Predators mingled with their prey in the desperate flight to safety. Many were trampled by the

---

407 Jonathan T. Hagstruma, Richard B. Firestone , Allen West, Zsolt Stefanka and Zsolt Revay. 2010. Journal of Siberian Federal University. Engineering & Technologies pages 123-132.

stampeding herd. And yet the worst was soon to come. In the distance, a towering wall of water raced toward them and engulfed them. Enormous trees snapped like twigs and floated as enormous rafts floundering in gigantic waves as the tsunami continued across the landscape. All but a few met their fate in minutes, drowned by the tidal wave. Those that survived the onslaught soon succumbed to their watery grave. As time progressed, piles of the dead were scattered across the Arctic.

The Ancients Cry Out: Tale of the Snake-head

The comet crossed the uppermost atmosphere of the Earth, bursting in air and casting fragments like artillery fire in all directions across four continents. Forests and grasslands were set on fire, sending dark black clouds of soot spiraling into the air, combining with an acrid cloud of gases and water vapor that swiftly surrounded the planet. And vicious storms erupted and pummeled the Earth with high winds, thrashing rains and hail, and enormous storm surges were followed by powerful tsunamis.

In November 1932, Edgar B. Howard archaeology research associate at the University of Pennsylvania Museum uncovered:

> *...the "matted masses of bones of mammoth." Mixed in with the bones were slender, finger-long spear points—Clovis points, as they are called today—which Howard carefully left in place. More than 10,000 Clovis points have been discovered, scattered in 1,500 locations throughout most of North America; Clovis points, or something similar, have turned up as far south as Venezuela. They seem to have materialized suddenly, by archaeological standards, and spread fast. The oldest securely dated points, discovered in Texas, trace back 13,500 years. In a few centuries they show up everywhere from Florida to Montana, from Pennsylvania to Washington State.[408]*

---

408 Mann, Charles C. 2013. The Clovis Point and the Discovery of America's First Culture. Smithsonian Magazine.

A strange mixture of fire, water, and ice had mercilessly ravaged the planet taking the lives of untold numbers of animals, plants and humans. *It is what I refer to as the 6ᵗʰ major extinction.*

The Younger Dryas and a Case for the Flood

A number of authors have speculated that the Younger Dryas and more recent glacial flooding events may provide a window into the Flood Account. For example, Ballard led a team that explored underwater archaeological sites in the Black Sea. Some suggest that the rapid flow of flood waters into the Black Sea points to Noah's Flood and the Babylonian Epic of Gilgamesh:

> *The Black Sea was once a freshwater lake, well below sea level. About <u>7,000 years ago</u>, according to geological evidence, the rising Mediterranean sea pushed a channel through what is now the Bosphorus, and then seawater poured in at about 200 times the volume of Niagara Falls. The Black Sea would have widened <u>at the rate of a mile a day, submerging the original shoreline under hundreds of feet of salty water</u>... There are many myths concerning a great flood in the region. There was a first mention in the Epic of Gilgamesh, the Babylonian work. The Romans and Greeks had the legend of Deucalion and Pyrrha, who <u>saved their children and animals by floating away in a giant box</u>. <u>The Hebrew book of Genesis most famously tells the story of Noah</u>... Dr Ballard began exploring the Black Sea in the Hull registered ship Northern Horizon, and used side-scanning sonar to look for interesting shapes on the seabed over a 200-sq-mile area, 12 miles off the Turkish coast, near Sinop.[409]*

The Black Sea and the Göbekli Tepe are near the Turkish home of Noah. Others continue the search in the same rock strata:

---

409 Radford, T. 2000. Evidence found of Noah's ark flood victims. theguardian.com

This article sheds new light on the narrative of Noah's Flood (Genesis Flood, Great Deluge) from a geoscientific point of view. It outlines the four most popular hypotheses: (i) the postglacial–early Holocene flooding of the Persian/Arabian Gulf which fell dry during the last glacial lowstand of the sea; (ii) a cosmic impact by a meteorite ca. 10,000 years ago, which triggered tsunami waves worldwide; (iii) the rapid re-filling of the Black Sea basin when the early Holocene rise of the Mediterranean Sea surpassed the Bosphorus sill about 8400 years ago; and (iv) the occurrence of one or several mega-floods in Central and Lower Mesopotamia, which left imprints in and around ancient settlement mounds (tells) such as Ur and Uruk.[410]

Without a doubt, the world has been through numerous flood cycles since the beginning of the Cambrian Explosion of Life. The bottom line is that the world has experienced catastrophic flood events, over and over again. Since the Cambrian there have been 6 major flood transgressions and 7 if you count the Younger Dryas flooding events. *The Younger Dryas Flood events are associated with the partial extinction of mankind and perhaps it is the prelude to the larger cycle that I call the 7th Apocalypse.*

The Younger Dryas Deep Freeze

Catastrophe upon catastrophe dominated the 6th Cycle. Asteroid impacts, incessant seismically induced plate tectonics, global flooding, mountain building, glaciation (referred to as the recent Ice Age), and mass extinctions were non-stop. Over the last several decades, scientists have been debating the cause of the Younger Dryas, an apparent abrupt period of sudden cooling during an otherwise warming trend. Whatever brought it about, also led to the end of the last "Ice Age." Rains from glacial impacts and fires led to at Tambora like year without summer. The impacts drove a mini Ice Age referred to as the Younger Dryas.

---

410 Brückner, H. and Max Engel. 2019. Noah's Flood—Probing an Ancient Narrative Using Geoscience. Palaeohydrology pp 135-151.

Isotope data from the GISP2 Greenland ice core suggests that Greenland was more than~10°C colder during the Younger Dryas and that the sudden warming of 10° ±4°C that ended the Younger Dryas occurred in only about 40 to 50 years…What can we learn from all this? The ice core isotope data were hugely significant because they showed that the Younger Dryas, as well as the other late Pleistocene warming and cooling events could not possibly have been caused by slow, Croll-Milankovitch orbital forcing, which occurs over many tens of thousands of years. The ice core isotope data thus essentially killed the Croll-Milankovitch theory as the cause of the Ice Ages.[411]

It shouldn't be surprising that the abrupt nature of the sudden drop of Younger Dryas temperatures is the consequence of multiple interrelated effects. Recent work on Greenland has led to the discovery of two impact sites beneath the Hiawatha Glacier.[412] However, the second crater appears to be slightly older based on stratigraphic studies. Some have hypothesized that a large asteroid or comet collided with the Laurentide Ice Sheet causing megafaunal extinctions like the wooly mammoths:

Firestone et al. (2007) proposed that an extraterrestrial object exploding over North America 12,900 years ago contributed to the megafaunal extinctions in North America and partially destabilized the Laurentide Ice Sheet and the thermohaline circulation in the northern Atlantic, thus triggering the Younger Dryas cooling event. The date of the event has been updated to 12,800 cal. BP by Kennett et al. (2015). In a subsequent paper with images of the Carolina Bays and the Nebraska Rainwater Basins, Firestone (2009) stated that the strikingly regular orientation of the bays was

---

411 Easterbrook, D.J. and A. Watts. 2012. The Intriguing Problem Of The Younger Dryas—What Does It Mean And What Caused It? wattsupwiththat.com
412 Vinas, M. 2019. NASA Finds Possible Second Impact Crater Under Greenland Ice. nasa.gov.

consistent with their formation by a shockwave coming from the Great Lakes. Firestone (2009), Firestone et al. (2010) also reported that impact material was found throughout the Carolina Bay sediments whereas these markers were found only in a thin layer elsewhere at the Younger Dryas Boundary; this can be interpreted as an indication that the bays are impact related.[413]

Perhaps a late Cenozoic barrage of asteroids or comets blasted the planet, melting enormous quantities of ice and triggering immense worldwide flooding. Enormous glacial Lake Agassiz is also believed to have released enormous quantities of flood waters across North America at this time.[414] Sound familiar? Evidence for impacts during the Younger Dryas period are being found at sites around the world from Greenland, Chile,[415] and Africa. A recent paper indicates 26 to 50 sites with evidences of impacts:

A team of scientists from South Africa has discovered evidence partially supporting a hypothesis that Earth was struck by a meteorite or asteroid 12,800 years ago, leading to global consequences including climate change, and contributing to the extinction of many species of large animals at the time of an episode called the Younger Dryas.[416]

---

413 Zamora, A. 2016. A model for the geomorphology of the Carolina Bays. Geomorphology. Volume 282. Pages 209-216.
414 Leydet, D.J. and others. 2018. Opening of glacial Lake Agassiz's eastern outlets by the start of the Younger Dryas cold period. Geology, Volume 46. Pages 155-158.
415 Pino, M. and others. 2019. Sedimentary record from Patagonia, southern Chile supports cosmic-impact triggering of biomass burning, climate change, and megafaunal extinctions at 12.8 ka. Scientific Reports. Volume 9
416 Thackeray, J. F. and others. 2019. The Younger Dryas interval at Wonderkrater (South Africa) in the context of a platinum anomaly. Palaeontologia Africana. Volume 54. Pages 30-35.

The 1ˢᵗ Extinction of Man

Worldwide impacts of such magnitude would result in extensive extinctions of life. During the Younger Dryas, the rock record indicates the extinction of millions of large animals including mammoths, mastodons, and saber-toothed tigers. And evidence links the Younger Dryas to the destruction of Clovis-age human populations.[417]

> *Clovis-age sites in North American are overlain by a thin, discrete layer with varying peak abundances of (i) magnetic grains with iridium, (ii) magnetic microspherules, (iii) charcoal, (iv) soot, (v) carbon spherules, (vi) glass-like carbon containing nanodiamonds, and (vii) fullerenes with ET helium, all of which are evidence for an ET impact and associated biomass burning at ≈12.9 ka. This layer also extends throughout at least 15 Carolina Bays... The shock wave, thermal pulse, and event-related environmental effects (e.g., extensive biomass burning and food limitations) contributed to end-Pleistocene megafaunal extinctions and adaptive shifts among Paleo-Americans in North America.[418]*

---

417 Moore, C.R. and others. 2017. Widespread platinum anomaly documented at the Younger Dryas onset in North American sedimentary sequences. Scientific Reports. Volume 7. Number 44031.
418 Firestone, R.B. and others. 2007. Evidence for an extraterrestrial impact 12,900 years ago that contributed to the megafaunal extinctions and the Younger Dryas cooling. PNAS. Volume 104. Issue 41. Pages 16016-16021.

19 *Evidence of the Younger Dryas Ice Age Apocalypse*

Out of the south cometh the whirlwind: and cold out of the north. By the breath of God frost is given: and the breadth of the waters is straitened.[419] Out of whose womb came the ice? and the hoary frost of heaven, who hath gendered it? The waters are hid as [with] a stone, and the face of the deep is frozen.[420] He casteth forth his ice like morsels: who can stand before his cold? He sendeth out his word, and melteth them: he causeth his wind to blow, [and] the waters flow.[421]

Each of the 6 Apocalyptic Cycles since the Cambrian Explosion of Life were accompanied by 6 periods of catastrophic global flooding and major extinction events. The Younger Dryas event appears to provide a short cycle of apocalyptic events triggered by an extraterrestrial impact. And the recent discovery of a Younger Dryas age impact on Greenland's Hiawatha glacier appears to provide the evidence sought for by many investigators. *Are the Younger Dryas events a prelude to the 7th Apocalypse or the close of the 6th Apocalypse?*

Platinum Fingerprint

The explosions of the Younger Dryas comet disintegrated as it entered Earth's atmosphere forming clouds of dust rich in the metal platinum (Pt) that coated the surface of the planet. An iridium (Ir) anomaly, associated with the Chicxulub impact, had

---

419 Job 37:8-10.
420 Job 38:29, 30.
421 Psalms 147:17, 18.

been the key marker for the demise of the dinosaurs at the beginning of the 6th Cycle. Iridium (Ir) is a member of the platinum group metals, so it is no surprise that platinum (Pt) would also be established as a marker for an impact layer near the end of the 6th Cycle or the triggering of the 7th Cycle.

Previously, a large platinum (Pt) anomaly was reported in the Greenland ice sheet at the Younger Dryas boundary (YDB) (12,800 Cal B.P.)... We document discovery of a distinct Pt anomaly spread widely across North America and dating to the Younger Dryas (YD) onset. The apparent synchroneity of this widespread YDB Pt anomaly is consistent with <u>*Greenland*</u> *Ice Sheet Project 2 (GISP2) data that indicated atmospheric input of platinum-rich dust... Of particular relevance to our study, Andronikov et al.23`24`25 investigated sediments from* <u>*Belgium, the Netherlands, Lithuania, and NW Russia near Finland*</u>*, reporting sharp YDB enrichment in Pt at the YD onset, as well as other meteoritic elements such as nickel, chromium, copper, and iridium. In a separate study, Andronikov et al.19 analyzed YDB magnetic microspherules from Blackwater Draw, New Mexico using scanning electron microscopy (SEM), electron probe microanalysis (EPMA), X-ray diffraction (XRD), and laser-ablation inductively coupled-plasma mass spectrometry (LA-ICP-MS). They reported an abundance peak in YDB microspherules... Results that are reported here for 11 sites from six U.S. states, from the Atlantic to Pacific coasts, provide strong evidence for above-background enrichment in Pt within sediments that date to the onset of YD climate...*[422]

Such a broad dispersion of the Pt layer around the planet, suggests that at least a portion of the comet exploded mid-air, much as the

---

422 Moore, C.R. and others. 2017. Widespread platinum anomaly documented at the Younger Dryas onset in North American sedimentary sequences. Scientific Reports. Volume 7, Article number: 44031.
422 Choi, Charles, Q. 2017. Comets: Facts About The 'Dirty Snowballs' of Space. Space.com

Tunguska asteroid of 1908 that flattened forests in Siberia. The Tunguska asteroid was estimated to burst in air at about 28,000 feet in elevation, traveling at about 34,000 miles per hour, and heating the air around it to about 44,000 degrees centigrade with an equivalent impact force of 185 Hiroshima bombs.[423] The Younger Dryas comet, would have had far, far more energy.

Bitter Waters and Rivers as Blood

Dry ice (carbon dioxide) when dropped into water will explode violently. Comets typically carry large amounts of frozen gases, including carbon dioxide and sulfur gases. The acids of these gases, from comets or volcanoes, quickly acidify water and turn it bitter, if it hadn't already been totally sublimated prior to impact.

> *The solid nucleus or core of a comet consists mostly of ice and dust coated with dark organic material, according to NASA, with the ice composed mainly of frozen water but perhaps other <u>frozen substances as well, such as ammonia, carbon dioxide, carbon monoxide and methane</u>. The nucleus may have a small rocky core. As a comet gets closer to the sun, the ice on the surface of the nucleus begins turning into gas, forming a cloud known as the coma. Radiation from the sun pushes dust particles away from the coma, forming a dust tail, while charged particles from the sun convert some of the comet's gases into ions, forming an ion tail... Scientists think short-period comets, also known as periodic comets, originate from a disk-shaped band of icy objects known as the Kuiper Belt beyond Neptune's orbit, with gravitational interactions with the outer planets dragging these bodies inward, where they become active comets. Long-period comets are thought to come from the nearly spherical Oort Cloud even further out, which get slung inward by the gravitational pull of passing stars. In 2017, scientists found <u>there may be seven times more big long-period comets than</u>*

423 NASA. 2008. The Tunguska Impact –100 Years Later. Science.nasa.gov.

previously thought...Some researchers are also concerned that comets may pose a threat to Earth as well.[424]

Meanwhile, the formation of iron carbonate and iron oxide which both take on a rust color, will make the rivers appear as blood. We see layers of these carbonates (siderite) and oxidized iron (hematite) in the banded iron formations of Northern Michigan.

The Rocks Cry Out: Pillars at Göbekli Tepe

Archaeologists now consider the *Göbekli Tepe* to be among the first, if not the world's first temple. These carved rocks found in Turkey, near the Syrian border, record an amazing account of a catastrophic encounter with a comet which has been correlated with the Younger Dryas.

By matching low-relief carvings on some of the pillars at Göbekli Tepe to star asterisms we find compelling evidence that the famous 'Vulture Stone' is a date stamp for 10950 BC ± 250 yrs, which corresponds closely to the proposed Younger Dryas event, estimated at 10890 BC. We also find evidence that a key function of Göbekli Tepe was to observe meteor showers and record cometary encounters. Indeed, the people of Göbekli Tepe appear to have had a special interest in the Taurid meteor stream, the same meteor stream that is proposed as responsible for the Younger-Dryas event. Is Göbekli Tepe the 'smoking gun' for the Younger-Dryas cometary encounter, and hence for coherent catastrophism?... What we can say is the following; • It is very likely that the people of GT had been keen astronomers for a very long time, and the low-relief carvings of animals (except snakes) symbolise specific asterisms. Pillar 43 very likely refers to the date 10,950 BC ± 250 yrs. • There is a consistent interpretation of much other symbolism at GT in terms of the YD event as a cometary encounter, which supports the theory of coherent catastrophism... Given the many threatening postures assumed by snake motifs at GT, the relationship snake/serpent = death and destruction is viable, but far from certain. Comets are certainly dangerous and destructive. Moreover, the serpent motif is a good symbolic representation of a meteor track... The headless man on pillar 43 indicates the event lead to loss of life[425]

Younger-Dryas Impact Driven Climate Reversal, a Worldwide Flood, and Mass Extinctions

The Younger Dryas was a time when giant buffalo, large zebras, and large wildebeasts went extinct in Africa. And human populations in Africa also plummeted based upon an abrupt end to the development of **Robberg Technology** represented by stone tools.[426] In North America, a long list of large mammals went extinct including numerous bird species, mammoths, wooly rhinoceros, horses, giant sloths, American lions, wolves, mastodons, giant armadillos to name but a few.

The Younger Dryas was the time of the most recent mass extinctions when 75% of the world's megafauna disappeared. The rapid melting of the North American ice sheet meant that global sea levels rose 400 feet... The Younger Dryas Extinction Rates aligns with the prehistoric end of the Ice Age... Human remains seem to decrease in the subsequent time periods. In the 3000-year period after the Younger Dryas, few archaeological records of humans are documented. This occurs between the abrupt ending of the Paleolithic era and the beginning of the Neolithic era.[427]

It appears that the Younger Dryas was associated with one or even several associated impact events that led to worldwide flooding, of

---

425 Sweatman, M.B. and D. Tsikritsis. 2017. Mediterranean Archaeology and Archaeometry, Vol. 17, No 1. Pages 233-250
426 Thackery, F. 2019. Did a large meteorite hit the earth 12,800 years ago? Here's new evidence. Phys.org.
427 Human Origin Project. Download 2020. Mass Extinctions of Prehistoric Ages. Humanorigin project.com

perhaps on a scale, with the similar flooding event on Mars. It is thought that impacts on the Laurentide ice sheet of North America were associated with the release of incredible volumes of violently released flood waters.

Oceanographers and marine geologists report what they call 'meltwater spikes' that occurred during the Younger Dryas. These are flows of fresh water from the ice cap into the oceans. There are two distinct glacial melt periods of prehistory.

- *Meltwater pulse 1-A (About 13 000 years ago)*
- *Meltwater pulse 1-B (About 11 600 years ago)*

Each coincides directly with the two extreme warming events that start and end the Younger Dryas. This evidence points to rapid sea level rises due to two ice cap melting instances. In this period the Clovis culture of North America was nearly wiped from the record, with only a few tribes surviving. Fossil records reveal a distinct layer of Clovis remains that date back to the Younger Dryas. The numbers then plummeted, presumed perished as the ice-sheet melted and temperatures changed. The Clovis culture was the dominant human population throughout North America at this time. Their sudden disappearance provides evidence that there may have been a cataclysmic cause of the Younger Dryas.[428]

The First "Extinction" of Man

Some believe that a massive impact occurred at the time of the Younger Dryas in the Antarctic Ocean (Southern Ocean).

A massive cosmic impact ~12,800 years before present in what is now the Southern Ocean delivered a catastrophic

---

428 Human Origin Project. 2019. Evidence that the Younger Dryas Forged Human Prehistory. Humanoriginproject.com

worldwide flood. Not long after the impact, the newly introduced waters flooded the Mediterranean Sea via the Strait of Gibraltar. The impact and its ensuing flood account for all reported Younger-Dryas effects; the worldwide flood and the Younger-Dryas event are synonymous. Culturally ubiquitous flood narratives corroborate the scientific record. Geology's "no flood, ever" paradigm is arguably the most profound error in the history of science, for it adversely affects geology, anthropology, archaeology, and matters concerning earth and early human history... the YD event is an episode marked by abrupt increases in snowfall and dramatic changes to flora, fauna, climate, and the oceans (Firestone et.al., 2007). Its precise cause is unknown, although it has been attributed by some to a cosmic impact roughly 12,800 years before present that has yet to be identified (Holliday, 2014; Wolbach et.al., 2018). The impact is reported to have induced YD effects across at least four continents (Kennett et.al., 2015), and it also formed an associated layer of nanodiamonds (Kennett et.al., 2009), microscopic diamond crystals that are created by very high-velocity collisions, found across most of the planet (Kinzie et.al., 2014).[429]

Likely, airbursts, smaller missile fragments and several large asteroid fragments were strewn across a wide global debris field from Europe to the Americas. Scientists are discovering that these impacts may be associated with volcanic eruptions in numerous places around the globe. The impact event could readily serve as the driver for both the pulses of glacial meltwater and volcanism. Some have correlated the eruption of the Laacher See volcano in Germany to the Younger Dryas, claiming that it would have set off an ash cloud across the northern hemisphere. It is part of the West European Rift system.

---

429 Jaye, Michael. 2019. The Flooding of the Mediterranean Basin at the Younger-Dryas Boundary. Mediterranean Archaeology and Archaeometry, Vol. 19, No 1, Pages 71-83.

Direct effects of the LSE included ash deposition, acid rain, wildfires, and increased precipitation, all of which could have affected the local and farfield ecology and cultures at the time (de Klerk et al., 2008; Baales et al., 2002; Engels et al., 2015, 2016)... Although the LSE preceded the most clearly expressed dynamical climatic change associated with the YD in central Europe, its timing (12.880 ± 0.040 ka BP) is indistinguishable from the Greenland temperature decrease...[430] *We find that a massive volcanic eruption occurred at 12,918 yr b2k and that immediately following the eruption the d-excess increases from 4 to 9 permil over a period of 37 years indicating a profound increase in sea-ice. During this time period, ratio of fluxes mantle to continental derived osmium also increases. Additionally, there is evidence of a 20-fold increase in extra-terrestrial osmium flux ~12,819 yr b2k following which the $\delta^{18}O$ values display a steep and sustained decline to −40 permil. These signals suggest that volcanism potentially induced the YD cooling, which may have been further exacerbated by an extra-terrestrial impact.*[431]

From my perspective, it appears that the impact of a comet and its multiple large fragments, could have caused the initiation of seismically induced volcanism and climate change, affecting the rerouting of ocean circulation. Perhaps a late Cenozoic barrage of asteroids and/or comets blasted the planet, melting enormous quantities of ice and triggering vast worldwide floods. The enormous glacial Lake Agassiz is also believed to have released enormous quantities of flood waters across North America at this time.[432] Sound familiar? Evidence for impacts during the

---

430 Baldini, J.U.L., and others. 2018. Evaluating the link between the sulfur-rich Laacher See volcanic eruption and the Younger Dryas climate anomaly. Climate of the Past. Volume 14. Pages 969–990.
431 Seo, Ji-Hye and Changhee Han. 2019. C11C-1305 - Younger Dryas Trigger Through the Lens of GRIP Ice Core. AGU Fall Meeting, San Francisco, CA. Poster Session.
432 Leydet, D.J. and others. 2018. Opening of glacial Lake Agassiz's eastern outlets by the start of the Younger Dryas cold period. Geology, Volume 46. Pages 155-158.

Younger Dryas period are being found in sites around the world from Greenland, Chile,[433] and Africa. A recent paper indicates 26 to 50 sites with evidences of impacts during the same period.

> *A team of scientists from South Africa has discovered evidence partially supporting a hypothesis that Earth was struck by a meteorite or asteroid 12,800 years ago, leading to global consequences including climate change, and contributing to the extinction of many species of large animals at the time of an episode called the Younger Dryas... The probability of a large asteroid striking Earth in the future may seem to be low, but there are thousands of large rocks distributed primarily between Jupiter and Mars. One in particular, classified as Apophis 99942, is referred to as a "Potentially Hazardous Asteroid." It is 340 meters wide and will come exceptionally close to the Earth in 10 years' time. "The closest encounter will take place precisely on Friday April 13, 2029," says Thackeray. "The probability of the Apophis 99942 asteroid hitting us then is only one in 100,000, but the probability of an impact may be even higher at some time in the future, as it comes close to Earth every 10 years."[434]*

Was the abrupt deep freeze during the Younger Dryas triggered by a sudden volcanic outburst that resulted from a meteoritic impact? Did this sudden volcanic episode lead to the buildup of sea ice that blocked the circulation of warmer Atlantic ocean currents from the poles? You be the judge.

> *The Younger Dryas (YD) abrupt cooling event (~12,900 yr to 11,600 yr) represents a brief return to severe cold conditions in mid- to high- latitudes in the Northern Hemisphere. The cooling is thought to have resulted from freshwater flooding*

433 Pino, M. and others. 2019. Sedimentary record from Patagonia, southern Chile supports cosmic-impact triggering of biomass burning, climate change, and megafaunal extinctions at 12.8 ka. Scientific Reports. Volume 9
434 Thackeray, J. F. and others. 2019. The Younger Dryas interval at Wonderkrater (South Africa) in the context of a platinum anomaly. Palaeontologia Africana. Volume 54. Pages 30-35.

of the northeast Atlantic and/or the Arctic Oceans that prevented deep water formation and promoted extensive southward expansion of sea-ice. Two different triggers that would lead to freshwater capping the north Atlantic have been proposed: (1) <u>catastrophic drainage of proglacial Lake Agassiz</u> and (2) <u>a meteorite impact-related partial destabilization and/or melting of the Laurentide ice sheet</u>... Recent revision in the age of Laacher See volcano (Volcanic Explosivity Index = 6) in Eifel, Germany has led to the suggestion that <u>the YD event was triggered by emplacement in the stratosphere of large amounts volcanic sulfur and halogens with sustained cooling resulting from a positive feedback involving sea ice expansion and/or AMOC shutdown.</u>[435]*

Based on the evidence, catastrophic events have profoundly shaped Planet Earth; uniformitarianism is but a secondary agent of change.

Revelations of Buttercup's Autopsy (Necropsy)

Among the herds of wooly mammoths, a large female, later named **Buttercup,** was tending to her young at her death, for she had given birth to eight calves based upon her autopsy.

Buttercup was found frozen, with her rear sticking upwards; it seems the mammoth was trapped in a bog, and tried to push down on her front legs to escape. But she couldn't. Trapped in the bog, she was exposed and defenseless, as predators moved in. Evidence on her skin indicates she was eaten. Eaten alive. Fortunately she must have sunk into the bog before too much of her was eaten, and her amazingly preserved body tells her story[436]

---

435 Seo, Ji-Hye and others. 2019. C11C-1305 - Younger Dryas Trigger Through the Lens of GRIP Ice Core. Abstract. AGU Fall Meeting, San Francisco, CA. AMOC = Atlantic meridional overturning circulation.
436 Freedman, J. 2014. Buttercup the Mammoth. Twilightbeasts.org

The autopsy showed that Buttercup's last meal was grasses and buttercups, hence her name. [437]

> *The mammoth's diet argues against the creature existing in a polar climate. How could the woolly mammoth sustain its vegetarian diet of hundreds of pounds of daily intake in an arctic region devoid of vegetation for most of the year? How could woolly mammoths find the gallons of water that they had to drink everyday? To make things worse, the woolly mammoth lived during the ice age, when temperatures were colder than today. Mammoths could not have survived the harsh northern Siberia climate of today, even less so 13,000 years ago when the Siberian climate should have been significantly colder. <u>The evidence above strongly suggests that the woolly mammoth was not a polar creature but a temperate one.</u> Consequently, at the beginning of the Younger Dryas, 13,000 years ago, Siberia was not an arctic region but a temperate one.* [438]

What caused the sudden drop in temperature so severe that mammoths, foxes, tigers, and mammoths have been found cryogenically preserved in the permafrost regions of the northern Siberian islands? Is the sudden drop in temperature connected to comet or asteroid impacts; perhaps a sudden downdraft of cold air from far up near the tropopause or even the mesopause? So we close the loop on Buttercup, and millions of others, some found with fragments of exploded meteorites in their tusks. They were frozen in a world on fire, with towering floods that eroded soil and rock from the land, piling it in a heap we now call the permafrost.

> *Though not as cataclysmic as the dinosaur-killing Chicxulub impact, which carved out a 200-kilometer-wide crater in Mexico about 66 million years ago, the Hiawatha impactor, too, may have left an imprint on the planet's history. The*

---

437 Ghose, T. 2014. Fresh Mammoth Carcass from Siberia Holds Many Secrets. Scientificamaerican.com.
438 Lescaudron, P. 2017. Of Flash Frozen Mammoths and Cosmic Catastrophes. Scott.net

timing is still up for debate, but some researchers on the discovery team believe the asteroid struck at a crucial moment: roughly 13,000 years ago, just as the world was thawing from the last ice age. <u>That would mean it crashed into Earth when mammoths and other megafauna were in decline and people were spreading across North America.</u> The impact would have been a spectacle for anyone within 500 kilometers. A white fireball four times larger and three times brighter than the sun would have streaked across the sky. If the object struck an ice sheet, it would have tunneled through to the bedrock, vaporizing water and stone alike in a flash. The resulting explosion packed the energy of 700 1-megaton nuclear bombs, and even an observer hundreds of kilometers away would have experienced a buffeting shock wave, a monstrous thunder-clap, and hurricane-force winds. Later, rock debris might have rained down on North America and Europe, and the released steam, a greenhouse gas, could have locally warmed Greenland, melting even more ice… The researchers proposed that besides changing the plumbing of the North Atlantic, the impact also ignited <u>wildfires across two continents that led to the extinction of large mammals and the disappearance of the mammoth-hunting Clovis people of North America.</u>[439]

Numerous recent studies confirm that there was a large impact at the time of the Younger Dryas period that had worldwide impacts including global flooding.

A massive cosmic impact ~12,800 years before present in what is now the Southern Ocean delivered a catastrophic worldwide flood. Not long after the impact, the newly introduced waters flooded the Mediterranean Sea via the Strait of Gibraltar. The impact and its ensuing flood account for all reported Younger-Dryas effects; the worldwide flood

---

439 Voosen, Paul. 2018. Massive crater under Greenland's ice points to climate-altering impact in the time of humans. Science. Sciencemag.com. AAAS.

and the Younger-Dryas event are synonymous. Culturally ubiquitous flood narratives corroborate the scientific record. Geology's "no flood, ever" paradigm is arguably the most profound error in the history of science, for it adversely affects geology, anthropology, archaeology, and matters concerning earth and early human history.[440]

Extensive analyses show that at least 23 sedimentary sequences on four continents support the synchronicity of the Younger Dryas. *This range overlaps that of a platinum peak recorded in the Greenland Ice Sheet and of the onset of the Younger Dryas climate episode in six key records, suggesting a causal connection between the impact event and the Younger Dryas.*[441]

The future of Planet Earth is uncertain. The population of the planet as of 2020 is roughly 7.8 billion with a projection that it will peak and stop growing by the end of 2100.[442] But many are concerned that humanity will be destroyed like the dinosaurs and so many other animals of the past. Nearly 40% of humanity lives within 100 kilometers of the coast at a time when coastlines are being threatened by rising seawater and massive storms. *In the U.S. nearly 5 million people live in 2.6 million homes at less than 4 feet above high tide – a level lower than the century flood line… and compounding this risk, scientists expect 2 to 7 more feet of sea level rise this century…*[443] *but this is just the expected beginning of the 7th cycle,* just as the Rock Record foretells. Fears are rising about world wars as nuclear arms are being accessed by more nations. Will mankind die by his own hands?

---

440 Jaye, M. 2019. The Flooding of the Mediterranean Basin at the Younger Dryas Boundary. Mediterranean Archaeology and Archaeometry, Vol. 19, No 1, Pages 71-83.

441 Kennett, J.P. and others. 2015. Bayesian chronological analyses consistent with synchronous age of 12,835–12,735 Cal B.P. for Younger Dryas boundary on four continents. PNAS.

442 Cilluffo, A. and N.G. Ruiz. 2019. World's population is projected to nearly stop growing by the end of the century. Pew Research Center.

443 Sealevel.climatecentral.org AND Kulp, S.A. and B.H. Strauss. 2019. New elevation data triple estimates of global vulnerability to sea-level rise and coastal flooding. Nature Communications **volume 10**, Article number: 4844.

Will Nature bring mass destruction to humanity in keeping with the projections of the Earth Chronicles? You be the judge.

The asteroid strike, officially verified in Greenland in November 2018, was first discovered in July 2015, but it took until November 2018 to confirm its source. According to NASA, the massive hole is "one of the 25 largest impact craters on Earth" and is said to have "rocked the Northern Hemisphere." Nearly two-thirds of all Americans want the government to focus on monitoring asteroids in the event of a catastrophic strike. One such possible asteroid that could cause this type of destruction is Apophis 99942, named for an Egyptian god of chaos.[444]

What is on the horizon for Planet Earth?

Apophis, god of chaos?

---

444 Ciaccia, C. 2019. Asteroid wiped out giant sloths, woolly mammoths 12,800 years ago. New York Post. nypost.com

20 *The 7ᵗʰ Apocalypse: Catastrophic Termination*

...there was a great earthquake; and the sun became black as sackcloth of hair, and the moon became as blood; And the stars of heaven fell unto the earth, ...And the <u>heaven departed as a scroll when it is rolled together</u>; and every mountain and island were moved out of their places. And the kings of the earth, and the great men, and the rich men, and the chief captains, and the mighty men, and every bondman, and every free man, hid themselves in the dens and in the rocks of the mountains...[445] for the first heaven and the first earth were passed away; and there was no more sea...[446]the heavens shall pass away with a great noise, and the elements shall melt with fervent heat, the earth also and the works that are therein shall be burned up.[447] But of that day and [that] hour knoweth no man...[448]

Planet Earth is miraculously suited for life. Whether you believe life originates from evolution, panspermia, planted by aliens, or Created; your very existence is a miracle. And whether you view Planet Earth in the context of the Universe, the Milky Way, or even the Solar system, it's as if Earth couldn't be better positioned to support your life. Yet scientists recognize that life is fragile and that the human race is vulnerable to extinction. Even the Scriptures say that at the end of time *there shall be a time of trouble such as never was since there was a nation*. Today astronomers fear that Earth will suffer a collision that will destroy humanity. Plans are even being made to transplant the human race to another world[449] before the inevitable happens. Humanity faces extinction by an asteroid impact just like the one that destroyed the

---

445 Revelation 6:13-15.
446 Revelation 21:1; 2 Peter 3:5-7.
447 2 Peter 3:9-12.
448 Mark 13:32.
449 Elon Musk. see SpaceX and the Mars Oasis.

dinosaurs; just like the one that led to the demise of the wooly mammoths during the Younger Dryas. Will Earth end by fire as it says, *for the first heaven and the first earth were passed away; and there was no more sea*.[450] This Scripture eerily echoes the past fate of Mars.[451] Could planet Earth suffer violence that could even cause loss of oceans? Or could nuclear reactors beneath our feet be at an explosion threshold that could release plasma, causing yet another cyclic outbreak of volcanism and plate tectonics?

An End-time Scenario: Could it Happen?

Panic seized the world's leaders as the Word of the coming destruction was verified. Scientists said it couldn't happen. They said that the probability was extremely low. Yet, there it was in the sky, the enormous object the size of a mountain was on a relentless path to collide with Earth. The earthly realms put their hopes and lives in the hands of the powerful. A League of Nations, a One World Order of sorts, drafted a multipronged initiative to save the Planet. The Space Force would take the lead to destroy or deflect the object, mustering every needed resource that mankind had developed, to defeat the celestial body. Government officials ordered that the information be withheld from the public to avoid mayhem.

The focus of planetary defense had been focused on protecting against smaller Near-Earth Objects (NEOs). Detection and assessment of the more elusive interstellar objects like *Oumuamua,* prophetically meaning *First Distant Messenger,* depended upon the astronomical Earth and Satellite observatories operated by the world's superpowers. These observatories were purposed for defense and war rather than on defense against rogue asteroids. However, several agencies charged with developing a means of defending against these threatening objects, had made progress that gave the Space Force hope.

---

450 Revelation 21:1; 2 Peter 3:5-7.
451 Robert I. Citron et al, (March 2018). Timing of oceans on Mars from shoreline deformation, Nature.

Astronomers and amateur stargazers watched as the odds of deflecting the interstellar object dwindled and hope turned to desperation. Would the preparation to deal with these near Earth objects be sufficient to defend against the enormous body heading to Earth? NASA had been rapidly accumulating information on these objects of potential doom. In 2004, NASA's Stardust program successfully retrieved samples from a comet designated as Wild-2. Unexpectedly, the mission returned puzzling material, much different from what scientists theorized. That wasn't good. Theory is one thing; reality can be something completely different. In 2005, NASA's Deep Impact spacecraft released an impactor on the surface of comet, Tempel 1. Subsequently, the European Space Agency's (ESA) Rosetta probe spent 2 years traveling with Comet 67P/Churyumov-Gerasimenko and finally made a controlled descent to its surface on September 30, 2016. Meanwhile, NASA was relying on the results of its impact test on the binary asteroid Didymos in 2022 to provide the data needed for the emerging strategy of attack on the enormous interstellar object threatening Earth. However, scientists realize that:

> *"To do something like this, we'd also need a really long warning time; the idea of a kinetic impactor is definitely not like [the movie] 'Armageddon,' where you go up at the last hour and you know, save the Earth," Chabot said. "This is something that you would do five, 10, 15, 20 years in advance — gently nudge the asteroid so it just sails merrily on its way and doesn't impact the Earth."* [452]

Mankind's Survival is in Imminent Danger

The 6th cycle began with the Chicxulub impact and ended before the Younger Dryas cold spell returned to its current warming trend. ***What is our fate now that the Younger Dryas event is past?*** We've crossed the precipice into the 7th Cycle and, if geologic history is an indicator, we now face invasions of floodwaters by

---

452 Bartels, M. 2019. Humanity Will Slam a Spacecraft into an Asteroid in a Few Years to Help Save Us All. Space.com.

sea and ***substantial catastrophic collateral events driven by a devastating nuclear explosion beneath our feet or a devastating impact from above***. We are far from prepared, yet the evidence indicates that world leaders are aware. But if we need 20 years to plan to deflect a massive extraterrestrial object, it might already be too late.

The National Atmospheric and Space Administration (NASA) has been rapidly preparing for potential devastating asteroid impacts. There are currently nearly 800,000 asteroids in our system and nearly 4,000 comets, of which less than 2000 are potentially hazardous asteroids (PHAs are larger than 100 meters in diameter). The next known significant asteroid flyby is in 2029 and 2036 when Apophis is projected to pass twice by Earth at about 19,000 miles with far less than a 1% chance of an impact.[453]

Simulation Scenario of a Rapidly Approaching Asteroid

NASA recently ran **a simulation** of a 60 meter asteroid fragment destined to impact New York City.

> *If this had been real, ten million people would be dead. NASA just tested our asteroid defense system, and despite eight years to plan their response, a team of scientists and engineers failed... The exercise began with an otherwise innocuous alert: an asteroid roughly 100 to 300 meters in size had been spotted in near-Earth orbit. According to early calculations, it had a one per cent chance of hitting the Earth on April 29, 2027. That's far more than the riskiest known real-world asteroid threat: Astronomers have calculated a rock called Bennu has a 1-in-2700 chance of striking Earth on September 25, 2135. A space craft is currently orbiting it, checking it out. But the hypothetical*

---

453 Nowakowski, T. 2017. Asteroid Apophis has one in 100,000 chance of hitting Earth, expert estimates. phys.org.

2019PDC asteroid bears a striking resemblance to another actual asteroid that is heading Earth's way - Apophis. It will also make its closest approach to Earth in April, 2029. Initially thought to have a 2.7 per cent chance of hitting our planet, refinements of its trajectory have all but ruled this out... [In the simulation] By 2021, NASA was able to get a probe off the ground to take a close look. This enabled calculations to be further refined. The prognosis was grim. It was predicted to strike above the US city of Denver... Finally, in August 2024, the 'impactors' were launched...They hit...But there was a problem. A large chunk of the asteroid had broken off. The scramble to figure out where it was headed was terrifying: New York. A race to launch military nuclear weapons to deflect this 60-meter projectile moving at 70,000 kilometres per hour was choked by political indecision. With just six months to go, the United States was left with just one choice. Evacuate. Prepare for the worst. Brace for impact. Naturally, things didn't go well...The simulated asteroid detonated 15km above New York's Central Park with the force of 1000 Hiroshima nuclear bombs. Everything within a 15km radius beneath it was simply obliterated. Nothing survived. Manhattan was gone. Structures as far out as 45km were shattered. Damage extended outward in a 68km radius.[454]

As you read through the simulation and study the details, you'll quickly recognize a couple of important factors. First, the fragment of the original object destined for New York City was only 60 meters in diameter, yet only asteroids greater than 100 meters are classified as PHAs. Yet its lesser mass was sufficient to wipe out a city the size of New York. Secondly, and perhaps more importantly, the analysis didn't consider the collateral damage that such an impact can cause. What about seismic wave effects along faults and fracture zones, including the potential activation of volcanic activity? What about global tsunamis and drownings of

---

454 Seidel, J. 2019. New simulation exercise aims to show NASA how people will really react when faced with asteroid threat. News Corp Australia Network.

coastlines? Are antipodal consequences being considered in the analyses? What about an impact on a large glacial ice mass like Greenland, the Arctic, or the Antarctic? Would an ice mass impact cause dangerous flooding along coastal cities? Would an ocean impact send tsunamis furiously racing towards coastal lowlands?

Asteroid Bennu, a 500 meter wide asteroid caused significant concerns when it was first discovered. It is expected to make its flyby around 2135. In response, the United States is developing a modular impactor design that could deflect such an asteroid.

> *Astronomers have calculated it has a 1-in-2700 chance of striking Earth on September 25, 2135...And if it strikes, it will unleash kinetic energy equivalent to that of 1450 megatons — some 80,000 times more powerful than the Hiroshima bomb. NASA's Centre for Near-Earth Object [NEO] Studies has a hit-list of 73 asteroids which have a 1 in 1600 chance of hitting the Earth. "Bennu was selected for our case study in part because it is the best-studied of the known NEOs," the researchers write. "It is also the destination of NASA's OSIRIS-Rex sample-return mission, which is, at the time of this writing, en route ..." The US government space agency and the National Nuclear Security Administration says its HAMMER (Hypervelocity Asteroid Mitigation Mission for Emergency Response) project could save life on Earth as we know it. Lawrence Livermore National Laboratory (LLNL) researchers have been busily calculating what it would take to budge the 79 billion kilogram lump of rock and ice.[455]*

Of course, scientists are concerned that blasting the asteroid might prove more fatal for life on Earth, since an impactor could break the asteroid into numerous fragments. These fragments in turn could produce a hit on the surface of the planet with a shotgun like blast, leaving a broad debris field of destruction, like that of the Younger Dryas.

---

455 Seidel, J. 2018. Asteroid Bennu is a threat to life on Earth. Now NASA has a plan to destroy it. News Corp Australia Network.

D.H. ALEXANDER

Galactic Superwaves: Cosmic Catastrophe?

NASA and the European Space Agency have confirmed a thesis proposed by Paul LaViolette of destructive superwaves,[456] fierce winds from a **supermassive black hole** that blow outward in all directions; a phenomenon that had been suspected, but difficult to prove until now. These winds of cosmic ray electrons, are moving outward from the galaxy's center at a third of the speed of light.

It is proposed that outbursts of cosmic ray electrons from the Galactic Center penetrate the Galaxy relatively undamped are able to have a major impact on the Solar System through their ability to vaporize and inject cometary material into the interplanetary environment. It is suggested that one such 'superwave', passing through the Solar System toward the end of the Last Ice Age was responsible for producing major changes in the Earth's climate and for indirectly precipitating the terminal Pleistocene extinction episode. The high concentration of ^{10}Be, NO_3^-, Ir and Ni observed in Lake Wisconsin polar ice are consistent with the scenario. The intensities of the Galactic nonthermal radio background and diffuse X-ray emission ridge are shown to vary with the Galactic longitude in the same manner as electron intensity along the proposed superwave 'event horizon'. The high luminosities and unusual structural features which characterize the Crab Nebula and Cassiopeia A are shown to be attributable to the fact that these remnants happen to coincide with this event horizon and are being externally impacted by an intense volley of relativistic electrons travelling from the Galactic Center direction. [457]

---

456 NASA. 2015. NASA, ESA Telescopes Give Shape to Furious Black Hole Winds.
457 LaViolette, Paul. 1986. Cosmic-Ray Volleys from the Galactic Center and their recent Impact on the Earth Environment. Earth, Moon, and Planets. Volume 37. Pages 241-286.

Could this phenomena have affected the Younger Dryas by injecting cometary material into the interplanetary environment? Will Planet Earth be confronted with other, yet to be discovered hazards, that could destroy our planet from unforeseen processes and events?

Plagues from thawing Permafrost

Plagues could emerge from bacteria and viruses, long frozen in the permafrost regions of the North where untold millions of dead animals are buried.

> *...viruses from the very first humans to populate the Arctic could emerge. We could even see viruses from long-extinct hominin species like Neanderthals and Denisovans, both of which settled in Siberia and were riddled with various viral diseases. Remains of Neanderthals from 30-40,000 years ago have been spotted in Russia. Human populations have lived there, sickened and died for thousands of years... In **a 2014 study**, a team led by Claverie revived two viruses that had been trapped in Siberian permafrost for 30,000 years. Known as Pithovirus sibericum and Mollivirus sibericum, they are both "giant viruses", because unlike most viruses they are so big they can be seen under a regular microscope. They were discovered 100ft underground in coastal tundra. Once they were revived, the viruses quickly became infectious...the study suggests that other viruses, which really could infect humans, might be revived in the same way.*[458]

There are many other potential catastrophic events that could be considered. Perhaps, two considerations remain. First, are we sure we have an accurate assessment of geologic time? Second, are the timing of prophetic events listed in the Scriptures firmly understood? Sir Isaac Newton spent as much time trying to decode the Scriptures as he did on science.

---

458 Fox-Skelly, J. 2017. There are diseases hidden in the ice, and they are waking up. bbc.com/earth.

21 *Time: A Matter of Perspective*

Be not ignorant of this one thing, that one day [is] with the Lord as a thousand years, and a thousand years as one day.[459] *Times are not hidden from the Almighty...*[460] *Remember the former things of old: for...God...Declaring the end from the beginning, and from ancient times [the things] that are not [yet] done...*[461]

Throughout this book, I have provided you with geologic time, reported from a very small slice of the enormous body of relevant literature. I have deliberately provided you with the references from the scientific community so that you can read the technical articles and make your own judgements. Dates, conventionally reported for the Cambrian Explosion of life, are greater than a half billion years. Like all information provided, the scientific community freely admits that there is uncertainty. In this chapter, I provide you with several considerations and uncertainties with respect to *time,* as measured by earth scientists. You now know that catastrophic events drive the formation of the features of the Earth that you see around you, from the mountains, to the continents, and ocean floors. Uniformitarianism is applicable during quiescent intervals of time when rates can be assured to be uniform. Rates of catastrophic processes are rapid. Yet, in either case the outcome can result in the same physical appearance. ***I STATE CONFIDENTLY: THE GEOLOGIC FEATURES OF OUR PLANET ARE DRIVEN BY CATASTROPHISM, NOT UNIFORMITARIANISM. LYELL'S PRINCIPLE AS APPLIED IS INCORRECT.***

---

459 2 Peter 3:8.
460 Job 24:1.
461 Isaiah 46:9, 10.

The Beginning of Time: The Big Bang was the Spoken WORD of God [462]

The Scriptures proclaim that *one day [is] with the Lord as a thousand years, and a thousand years as one day.*[463] *Times are not hidden from the Almighty...*[464] *A thousand years in thy sight are but as yesterday when it is past...*[465] Both science and the Scriptures agree that the universe had an instantaneous beginning, or *singularity*. Science contends that the Universe originated as a Big Bang from nothing. The Scriptures assert that *by Him all things were created that are in heaven and that are on earth, visible and invisible* and *the heavens were stretched out as a curtain.*[466] And *the key word is STRETCHED*. God stretched the Universe into existence in a manner that astrophysicists reconstruct. Astrophysicist Hugh Ross[467] and physicist Gerald Schroeder view time from **the perspective of two viewers: God outside of time, and man within the space-time continuum. Without a doubt, my eternal God stands outside the Universe, independent of time and space, as the Creator of the Universe and everything that is in it, including the Creation of Earth and Life as described in Genesis. I also accept that the Bible is the inerrant Word of God as I will prove in a subsequent series of books.**

Dr. Schroeder has published an evaluation concluding that the two perspectives are equivalent yet viewed from two vantage points: God's vantage point is outside the constraints of space and time; man's vantage point is constrained within the framework of space-time.[468] **His explanation is as follows:**

---

462 Psalms 33:9.
463 2 Peter 3:8.
464 Job 24:1.
465 Psalm 90:4.
466 Isaiah 40:22; 45:12.
467 Ross, Hugh. 2004. A Matter of Days, Resolving a Creation Controversy. Colorado Springs: NavPress. ISBN 978-1576833759
468 Schroeder, G. 2013. Age of the Universe. Geraldschroeder.com WordPress and Aish.com.

From the perspective of science, the Universe is on the order of 13,900,000,000 years old whereas from the perspective of God the age of the Creation is a mere ~6,000 years. Perhaps, time is a matter of perspective. An observer outside the universe would have one perspective and observers within the Universe would have a much different perspective. The stretching of space-time is the central consideration.

Einstein, in the laws of relativity, taught the world that time passes at different rates in different environments. Absolute time does not exist in our universe. The passage of time is relative. In regions of high velocity or high gravity, time actually passes more slowly relative to regions of lower gravity or lower velocity. (One location relative to another, hence the name, the laws of relativity) This is now proven fact. Time actually stretches out... The Bible views time looking forward into the expanding space of the universe from the moment of the threshold rest energy of a proton, a moment that was a tiny fraction of a second following the big bang creation of the universe, when the universe was vastly smaller than it is today. ...The earth based scientific measure views time looking back in time from the present toward the threshold rest energy of a proton, a moment that was a tiny fraction of a second following the big bang creation of the universe, a time when the universe was vastly smaller than it is today.

Space stretches, and that stretching of space totally changes the perception of time... Let's look at the development of time, day-by-day, based on the expansion factor. Every time the universe doubles, the perception of time is cut in half. Now when the universe was small, it was doubling very rapidly. But as the universe gets bigger, the doubling time gets longer. This rate of expansion is quoted in "The Principles of Physical Cosmology," a textbook that is used literally around the world... using the exact expansion factor (the ratio of the energy or temperature of space at the threshold energy of protons [10.9×10^{12} K] to the ...corrected current energy or

temperature of space [3.03 K]). The result gives an overall age of the universe of 13.9 billion years:

Day 1: The first of the Biblical days lasted 24 hours, viewed from the "beginning of time perspective." But the duration from our perspective was 7 billion years.
Day 2: The second day, from the Bible's perspective lasted 24 hours. From our perspective it lasted half of the previous day, 3.5 billion years.
Day 3: The third 24 hour day also included half of the previous day, 1.8 billion years.
Day 4: The fourth 24 hour day — 0.9 billion years.
Day 5: The fifth 24 hour day — 0.5 billion years.
Day 6: The sixth 24 hour day — 0.2 billion years.

The arguments of Ross, Schroeder and numerous others is essentially the view of "Old Earth Creationists." Can a case be made for "Young Earth Creationists"? You be the judge.

Missoula Flood: Catastrophism

In 1925, Bretz concluded that the Columbia Basin had been the site of an enormous catastrophic flood. His findings set off a firestorm in the scientific community because his conclusion contradicted the widely accepted model of uniformitarianism set forth by Charles Lyell. The *"heretical"* theory[469] proposed by Bretz was hotly contested, not only because it challenged the theory of uniformitarianism, but because it appeared to support the *Biblical Flood* and was called an *outrageous hypothesis.*[470] But in 1972, NASA and the Geological Survey released satellite imagery of the Columbia Basin, vindicating the theory put forward by Bretz. *What appeared to take millions of years to erode the Columbia Basin, took days.* Once again, catastrophism defied geologic time.

---

469 Bretz, J Harlen (1925). "The Spokane flood beyond the Channeled Scablands". *Journal of Geology.* **33** (2): 97–115, 236–259.
470 Cassandra Tate, Bretzm J Harlen (1882-1981), Geologist. HistoryLink.org Essay 8382. Posted November 27, 2007.

8:32AM May 18, 1980 Eruption of Mount St. Helens, Skamania County, Washington, U.S.

"Vancouver! Vancouver! This is it!" Those words, spoken by Dr. David Alexander Johnston, will echo through history as he sounded the warning at 8:32 AM on Sunday morning May 18, 1980. The bulge that had been forming on the side of the mountain for the preceding months catastrophically exploded, killing David. The explosive power was in excess of 30,000 Hiroshima size atomic bombs. Within minutes majestic stands of trees were blown away. Yet, out of the disaster, *Nature provided a lesson book in the aftermath of the explosion that undermines the theory of uniformitarianism.* The catastrophic explosions and the physical and chemical processes that it set in motion, brought about geologic features that are typically assessed by uniformitarian processes to take millions of years. Over the years following the initial blast, a canyon greater than 100 feet deep was formed, and aptly named the "Little Grand Canyon." Initial age dating of the newly formed dome rock gave *anomalous dates for the whole rock of 350,000 years and for the minerals it contained from 340,000 to 2.8 million years even though the sample was only about ten years old.*[471] These dates underscore the controversy of age dating, especially with respect to the mixing of older material, or chunks of rocks (xenoliths), with younger magmatic fluids. The catastrophic events at Mount St. Helens led to the rapid formation of strata and laminated structures that could be misinterpreted as taking thousands, even millions of years to form. In Spirit Lake, the deposition of numerous trees cross numerous layered strata (poly strata or polystrate fossils) demonstrate that the enclosing rock layers formed rapidly due to a catastrophic event that caused burial by the rapid layering of sediments. These polystrate fossils show that the transected layers were laid down in days undermining conventional wisdom of thousands to millions of years for each layer. Finally, zones of

---

471 Austin, S.A., "Excess Argon Within Mineral Concentrates from the New Dacite Lava Dome at Mount St Helens Volcano," *CEN Tech. J.* **10**(3):335–343, 1996.

rapid erosion (including Canyon formation) and even striated rock surfaces (normally attributed to glaciation) are evident at Mount St. Helens. **The point is, that one could easily be misled by the origin of the features and the duration and magnitude of the accompanying events. The same features that are usually attributed to thousands or even millions of years could form within a short period of time, in this case in days to decades.**

Turbidite Layers and Sedimentation Rates

On November 18, 1929 an earthquake triggered a mudslide off the coast of Newfoundland. As fortune would have it, we had an opportunity to measure the velocity of that mudslide as it raced down the continental slope to the depths of the ocean floor and made its way towards the coast of Portugal. As it travelled across the ocean it severed 12 trans-Atlantic cables. These cables served as stop watches, providing the data to determine that the mud flow, referred to as a turbidite, was traveling upwards of a 100 kilometers an hour. Why is this significant? What it means is that a single stratigraphic rock formation can be deposited over hundreds of kilometers in hours or days, rather than upwards of millions of years. *We know that numerous turbidite layers are found throughout the sedimentary rocks since the Cambrian.* Similarly, numerous studies indicate that the sedimentological rate can be highly variable with rates that are far faster than the rates typically assigned to the stratigraphic units. As you'll recall:

> *Based on the sedimentation analysis of the COS [Cambrian-Ordovician Sandstone Sequence] from the Leningrad district, "pure" sedimentation time for Lower Paleozoic sands can be estimated at 100–200 yr. The paradox is that geological time of the Sablinka sequence formation amounts to 10–20 Ma (Tugarova et al., 2001, p. 89). The authors explain this paradox by the rewashing of sediments in shallow water marine conditions with active lithodynamics, where processes of accumulation and seafloor erosion occur*

side by side and replace one another depending on parameters of storms and currents.[472]

Rapid Preservation and Transitional Ecosystems

The **Cambrian Explosion** of life is based upon exceedingly well preserved marine fossils. The preservation of these fossils are almost universally associated with **catastrophic turbidite beds**. The living animals are engulfed in a sudden cloud of mud that suffocates them and preserves them. But this mechanism is not limited to marine fossils from trilobites to whales. The same holds true for reptiles, dinosaurs, and mammals. Had the animals died a natural death, their body parts would rapidly disintegrate or be eaten by predators. A uniformitarian burial mechanism would not be sufficient for preservation. **Most were catastrophically buried. And it appears that whole Transitional Ecosystems were buried.**

Rock Formations by Differing Mechanisms: Salt

Scientists estimate that it will take nearly 1 billion years to evaporate our current oceans.[473] If current oceans were totally evaporated, they would leave a layer of salt 500 feet thick[474] over the surface of the Earth. If one were to sum the actual thicknesses of evaporated salt deposits on the continental United States[475] one can come up with over 6000 feet of salt. Could it be that the majority of massive salt deposits on planet Earth come from a mechanism other than solar evaporation? As mentioned earlier, **salt deposits show a high correlation with tectonic zones,**

---

472 Berthault, G. and others. 2009. Reconstruction of Paleolithodynamic Formation Conditions of Cambrian–Ordovician Sandstones in the Northwestern Russian Platform. Lithology and Mineral Resources, Vol. 46, No. 1, pp. 60–70. © Pleiades Publishing, Inc.

473 Leconte, J. and others. 2013. Increased insolation threshold for runaway greenhouse processes on Earth-like planets. Nature. Volume 504. pp. 268-271. ALSO Science News. 2013. When will Earth lose its oceans? Sciencedaily.com.

474 https://www.usgs.gov/special-topic/water-science-school

475 Pierce, W.G. and E.I. Rich. 1962. Summary of Rock Salt Deposits in the United States as Possible Storage Sites for Radioactive Waste Materials. Geological Survey Bulletin 1148.

especially rift zones. Rift basins throughout the Middle East are sites of Cambrian salt precipitation in subsiding rift environments along the Middle Eastern edge of Gondwanaland.[476]

Extremely low solubility of typical seawater salts within certain supercritical sections of their pressure-temperature composition space is a proven experimental fact... <u>Our alternative model for the formation of salt deposits hypothesizes that high temperatures and pressures characteristic for the high heat- flow zones of tectonically active basins may bring submarine brines into the out-salting regions and result in the accumulation of geological-scale salt depositions.</u>[477]

I conclude that <u>*massive geologic salt deposits are the product of precipitation, serpentinization, and supercritical precipitation of salt in tectonic zones as evidenced by Dead Sea Salts*</u>.

Newton's Captivation with Prophetic Time

We couldn't end this chapter without briefly mentioning prophetic time. Sir Isaac Newton's book on prophecy, ***Observations upon the Prophecies of Daniel, and the Apocalypse of St. John***[478] reflects his deep interest in the supernatural. *Newton was stunned by the accuracy of the time prophecies of Daniel 2:31-35. He wondered how Daniel, living in the time of Babylon, could accurately name the coming kingdoms of Persia, Greece, Rome, and the modern era in advance? Especially, since Greece was explicitly named nearly 200 years after the death of Daniel.*[479] Daniel warns that the world will end with a

---

476 Moujahed I. Husseini and Sadad I. Husseini. 1990. *Origin of the Infracambrian Salt Basins of the Middle East.* Geological Society, London, Special Publications, 50, 279-292.
477 Hovland, M. and others. 2006. Basin Research. Volume 18. Pp. 221-230.
478 Newton, Isaac, Sir. Observations upon the Prophecies of Daniel, and the Apocalypse of St. John. London. J. Darby and T. Browne. 1733.
479 Daniel 8:21.

Stone cut without hands.[480] As a result, *the heavens shall pass away with a great noise, and the elements shall melt with fervent heat, the earth also and the works that are therein shall be burned up.*[481] The following timeline is from Daniel who lived about 600 B.C.:

World History in Advance Daniel 2:31-35
Head of Gold **Babylon** 605-539 BC
Arms of Silver **Persia** 539-331 BC
Belly, Thighs of Brass **Greece** 331-168 BC
Legs of Iron **Rome** 168 BC-476AD
Feet of Iron, Clay **Modern Day** 476AD - Present
Stone Cut without Hands **Near Future** **Last Trump**

Today, the future of the human race is in great danger. Will the world be ready? Will you be ready? You be the judge.

---

480 Daniel 2:34.
481 2 Peter 3:10.

EPILOGUE

Multitudes, multitudes in the valley of decision: for the day of the LORD [is] near in the valley of decision. The sun and the moon shall be darkened, and the stars shall withdraw their shining.[482]

Catastrophism is responsible for resurfacing the Face of Planet Earth; not uniformitarianism. The Face of the Earth has changed catastrophically during each of 6 near-common cycles since the Cambrian. Pattern analysis shows that cycles of plate tectonics are driven by sudden periodic injections of energy within the mantle due to impacts and/or nuclear explosions. These injections of energy cause heating that lift ocean floors, forcing seawater to invade dry land. During the 6th Apocalypse, in just the last 1.5% of geologic time, vast new ocean floors have been paved. This continuum of Catastrophic Cycles challenges the very foundation of evolution. All signs are pointing to a 7th Apocalyptic cycle that is here NOW. It appears that the Earth Chronicles and Scriptural Prophecy are both pointing to the end of time. If not evolution, then what is the origination of life? Could there be an Intelligent Designer after all?

---

482 Joel 3:14, 15.

22 Conclusions: Catastrophism and Intelligent Design

The words [were] closed up and sealed till the time of the end,[483] *the mystery which hath been hid from ages and from generations, but now is made manifest.* [484] *There shall come in the [1] last days scoffers… saying... [2] all things continue as [they were] from the beginning of the [3] creation… they willingly are ignorant… that by the Word of God [4] the heavens were of old, and [5] the earth standing out of the water and in the water: [6]the world that then was, being overflowed with water, perished: the heavens and the earth, which are now, by the same word are kept in store, [7] reserved unto fire...*[485]

Somewhere, somehow, in an instant of time, life erupted on planet Earth. Science cannot explain how, when, or why it happened. None have been able to pinpoint the step by step chemical processes and biochemical mechanisms that led to the first living cell. Science, at present cannot tell you how the first living cells came to be; science can only tell you that it wasn't by accident. Why? Because the creation of a living animal cell from non-living constituents is amazingly and incredibly too unlikely to have happened by accident. How was information encoded within the complex proteins of the first living cell? How were these cells able to reproduce? How did a myriad of different eyes form

---

483 Daniel 12:9.
484 Colossians 1:26.
485 2 Peter 3:3-7.

without predecessors? How is it that these first eyes also happened to be supported by the appearance of the brain? Because of such complexity, many scientists avoid investigations of the initial complex steps of the origin of life. They bypass these steps with a *leap of faith*. Without a confirmation or a convincing explanation of such early steps, evolution is nothing but a concept at best. *The alternative explanations to evolution is that life was planted on Earth; either by Creation, aliens, or panspermia.*

Scientists like Johannes Kepler (1571-1630) and Sir Isaac Newton (1643-1727) derived elegant mathematical equations and fundamental physics to describe the universe. This elegance cemented their view that the universe originated from a designer God. They saw numerous examples of *Intelligent Design* and *Fine Tuning* that many astronomers and physicists[486, 487] still point to today. During the 1700's and the 1800's their thinking was largely cast aside. *Workers in the emerging field of geology soon concluded that the origin of the Earth was incredibly old. These new geologists extrapolated the age of the Earth by applying the exceedingly slow rates of processes we see today. The resulting time spans appeared to undermine the Biblical origin of the world. But because the geologic section is comprised of catastrophic rates and processes; times since the basal Great Unconformity need to be revisited.*

CONCLUSION 1: 6 Cycles of Past Catastrophes

The *Earth Chronicles* clearly document 6 major cycles of catastrophe, beginning with the *Cambrian Apocalypse*. Each of these 6 cycles is readily defined by sedimentary rocks that give evidence of periods of continental flooding (transgression and regression of seas upon the land). The transgressions typically accompany a period of active plate tectonics. It is possible that

---

486 Guillermo Gonzalez and Jay W. Richards. 2004. *The Privileged Planet: How Our Place in the Cosmos Is Designed for Discovery.* Simon and Shuster. 464 pp.

487 L. Susskind (2005). *The Cosmic Landscape: String Theory and the Illusion of Intelligent Design.* Little, Brown.

nuclear explosions at the core-mantle boundary serve as the triggers for these Apocalyptic Cycles. In a number of Apocalyptic cycles, *extraterrestrial impacts* are undoubtedly correlated with plate tectonics and extinction. The Chicxulub impact that triggered the *Cenozoic Apocalypse* (the 6th Cycle) is a clear example. Impact events are often defined by a layer of soot, platinum group metal dust, and glass spherules among other evidence. If severe enough, the impact sends seismic waves through the planet that cause volcanic activity. Volcanic eruptions can occur *antipodally* and collaterally to the impact. The ocean ridges, as seen on the back cover, rise up as if swollen (as would be the case of a nuclear explosion). Their mountainous rise, displaces enormous amounts of sea water which is forced to spill catastrophically across the continents, eroding the land and causing inland seas. The collateral flooding causes turbidite mud flows that race along the margins of continents and inland seas, resulting in the ***burial and extinction of marine life.*** Greater than 95% of extinctions are attributed to marine life. Collateral effects often include erosion, mountain building, klippes, subduction of old oceans, glacial flooding, and volcanic eruption that creates new ocean crust.

CONCLUSION 2: Fountains of the Great Deep, Subsurface Oceans, and Global Flooding

Multiple episodes of undisputed global flooding are documented in the *Earth Chronicles* since the beginning of the *Cambrian Explosion of Life*. In some instances, impacts on continental glaciers appear to have initiated catastrophic flooding as evidenced on Mars. There appear to be clear markers of this event sequence during the Younger Dryas. Scientific investigations reveal a *deep mantle "ocean reservoir" of water* based on mineralogical studies of ringwoodite and diamonds. These minerals, lock ocean water into their structures. It appears that a high pressure form of Ice VII is also prevalent in the mantle. *Subduction of ocean floor appears to serve as a conveyor belt that circulates surface seawater to the mantle and back to the surface.* Studies of Saturn's moon, Enceladus proves that jets of water, like *Biblical fountains of the great deep*, can be released under great pressure and heat from the interior of a moon or planet. It would not be surprising that such fountains of the great deep were released from deep tectonic lineaments such as grabens and mid-ocean ridges during episodes of catastrophic Flooding. Four lines of evidence of these cycles of global flooding are provided below:

First Line of Evidence: an incomparable global erosion[488] event is found at the Precambrian-Cambrian boundary, referred to as the *Great Unconformity*. Scientists believe that both glaciation and extensive flooding are the causes for this worldwide feature.

Second Line of Evidence: Six worldwide mega periods of global flooding based upon 6 worldwide invasions of land by the seas, referred to as the Sauk, Tippecanoe, Kaskasia, Abseroka, Zuni, and Tejas Transgressions are recorded in sedimentary sequences. The *Sauk marine sediment sequence* rests on the Great Unconformity at the base of the Grand Canyon sedimentary column.

Third Line of Evidence: As molten magmas rise through the mantle and into the crust of the oceans, their _**buoyancy raises crustal oceanic spreading centers[489] and causes the displacement of ocean waters onto the continents.**_ Imagine the thousands of miles of mid-ocean volcanic chains being lifted along the sea floor all at once. Their sheer volumes and associated earthquakes would displace water onto the continents. *The Cambrian CO_2 spike attests to the enormity of worldwide volcanic eruptions that created new ocean floors*. In some instances, high sea levels occurred concurrently with relatively low lying continental terrain.

Fourth Line of Evidence: Layer upon layer of continent-wide stratigraphic facies with constant compositions and thicknesses gives additional evidence of global flooding. Flow patterns suggest that massive quantities of sediments were transported by surging walls of water across the continents as evidenced by turbidite beds. Today extensive stratigraphic units, arrayed like a layer cake, cover enormous areas of the Earth's surface. *The*

---

488 Keller, C.B. and others. 2018. Neoproterozoic glacial origin of the Great Unconformity. PNAS. Volume 116. Number 4.
489 National Geographic. 2015. Seafloor spreading. nationalgeographic.org.

Tapeats sandstone covers much of North America and is correlated with analogous geologic sequences in North Africa, Eastern Greenland, Israel, South Australia, and portions of South America.

CONCLUSION 3: Earth's Resurfacing is by Catastrophism; Not Uniformitarianism

During the 6 Cycles of Catastrophe, the face of the planet has changed many times. At the beginning of the Cambrian, the supercontinent Rodinia, broke into fragments, which later assembled into a supercontinent known as Pangea, which finally broke apart taking on the configuration of today's continents. Each of the 6 Cycles we've discussed, are born of a series of intertwined catastrophic events commonly initiated by a significant crust busting event that radiates vibratory seismic waves through the planet. These seismic waves shake the mantle causing crustal fractures, doming, flexure, and buoyant uplift due to the upwelling of magmas along immense crustal lineaments. In the 6th cycle, the Rock Record bears witness to the resurfacing of the Planet by the outpouring of incredible volumes of upwelling magmas, largely within less than 1.5% of Earth's history according to geologic time. As time progresses, this outpouring can form whole new oceans along Mid-Ocean Ridges (MOR) and displace and eventually subduct and bury former oceanic crust beneath the less dense silica rich continental rafts. Consequently, as these upwelling mantle-magmas buoy up the ocean floors, enormous volumes of seawater is displaced from the oceans, forcing flood waters to race inland, even over continents. Given sufficient impact energy, the collisions among continents result in mountain-building. The high energy of these events cause muds to race downward from upland shores and continental shelves into the depths of oceans and seas. And in so doing, the fine grained muds envelope marine animals suffocating them in a turbulent cloud, instantly burying plants and animals, big and small. Therefore, the soft bodied Cambrian marine creatures and even their delicate organs are preserved.

CONCLUSION 4: A Case for Impact and/or Nuclear Explosion driven Plate Tectonics

Several authors have noted that their data suggests: 1) the continents moved at rapid rates and 2) sea floor production appears anomalously high. Some workers note that the earlier production of the Cenozoic sea floor was significantly more rapid than the rates in the younger Cenozoic.

> *... rapid changes during the Cenozoic defy models of steady-state seafloor formation, and demonstrate the time-dependent and evolving nature of plate tectonics on Earth.[490] (1). The sudden appearance of virtually all the animal phyla (2) and their exponential diversification are coeval with abrupt shifts in oceanic geochemistry (3, 4). Recent calibration of this time interval with U-Pb isotopic ages (5, 6) indicates that these events occurred within a span of 30 million years (My), and the major diversification happened in only 10 to 15 My... The new ages, along with paleomagnetic data, indicate that <u>continents moved at rapid rates</u> that are difficult to reconcile with our present understanding of mantle dynamics (7). [491]*

From my perspective, this suggests that the Chicxulub impact triggering event, at the beginning of the Cenozoic, imparted significant, sudden energy to the mantle. The result was rapid formation of new seafloor as plates moved apart. As this energy dissipated, the rates of sea floor production decreased to present day rates. It is amazing that the largest percentage of the current ocean floors were rapidly extruded as magma and converted to rock, all in less than 1.5% of geologic time as measured by geologists. Frictional forces are associated with rapid subduction of ocean crust diving into the mantle near the edges of continents

490 Conrad, C.P. and C. Lithgow-Bertelloni. 2007. Faster seafloor spreading and lithosphere production during the mid-Cenozoic. Geology Volume 35(1).
491 Kirschvink, Joseph, L. Robert L. Ripperdan, and David A. Evans. July 1997. Evidence for a Large-Scale Reorganization of Early Continental Masses by Inertial Interchange True Polar Wander. SCIENCE. Vol. 277. Pp. 541-545.

and island arcs. Yet subduction is lubricated by supercritical sea water, salts, and carbon dioxide released from serpentinized peridotite. Stern and others *suggest that a thin channel decouples the lithosphere from the asthenosphere and allows plate tectonics to take place.* [492] *It is feasible that a nuclear explosion at Earth's core–mantle boundary (CMB), could cause shock waves to propagate through the Earth... A shock wave created by rapidly expanding plasma resulting from such an explosion could disrupt, warp, fracture and heave overlying mantle and crust material. Given a number of factors, it appears that the 6 cycles could be associated with cyclical nuclear explosions at the Core-Mantle boundary. (Fukuhara, 2016 and de Meijer, 2013).*

CONCLUSION 5: Salt Layers can Out-Salt from Supercritical Fluids by Tectonic Forces

Rift basins throughout the Middle East are sites of Cambrian salt precipitation in subsiding rift environments along the Middle Eastern edge of Gondwanaland.[493] Massive geologic salt deposits, at least in some locations, appear to be the product of supercritical precipitation of salt in tectonic zones. The circulation of salt-rich seawater in rapidly subducting slabs and along rift valleys provides an environment for such extensive salt deposits.

Extremely low solubility of typical seawater salts within certain supercritical sections of their pressure-temperature composition space is a proven experimental fact. Its consequences are often referred to as either 'shock crystallization' or 'out-salting'. Our alternative model for the formation of salt deposits hypothesizes that high temperatures and pressures characteristic for the high heat-

---

492 Stern, T.A. and others. 2015. A Seismic Reflection Image for the Base of a Tectonic Plate. Nature volume518, pages85–88

493 Moujahed I. Husseini and Sadad I. Husseini. 1990. *Origin of the Infracambrian Salt Basins of the Middle East.* Geological Society, London, Special Publications, 50, 279-292.

£ow zones of tectonically active basins may bring submarine brines into the out- salting regions and result in the accumulation of geological- scale salt depositions.[494]

CONCLUSION 6: Catastrophism Undermines the Foundation of Evolution

Darwin, raised four primary concerns about his own hypotheses of the origin of life on Earth:[495]

Firstly, why, if species have descended from other species by insensibly fine gradations, do we not everywhere see innumerable transitional forms?...why do we not find them embedded in countless numbers in the crust of the earth?

Secondly, is it possible that an animal having, for instance, the structure and habits of a bat, could have been formed by the modification of some animal with wholly different habits? Could we believe that natural selection could produce, on the one hand, organs of trifling importance such as the tail of the giraffe, which serves as a fly-flapper, and on the other hand, organs of such wonderful structure, as the eye, of which we hardly as yet fully understand the inimitable perfection.

Thirdly, can instincts be acquired and modified through natural selection? What shall we say to so marvelous an instinct as that which leads a bee to make cells, which have practically anticipated the discoveries of profound mathematics?

Fourthly, how can we account for species, when crossed, being sterile and producing sterile offspring, whereas, when varieties are crossed, their fertility is unimpaired?

---

494 Hovland, M. and others. 2006. Basin Research. Volume 18. Pp. 221-230.
495 Darwin, Charles, 1809-1882. (1859). Chapter 6. Difficulties on Theory. On the origin of species by means of natural selection, or preservation of favoured races in the struggle for life. London: John Murray.

More than 150 years have elapsed since Darwin's manuscript, *On the origin of species by means of natural selection, or preservation of favoured races in the struggle for life*, was published, yet no new significant evidence has been brought forward that answers the crippling weaknesses of evolution raised by Darwin himself. No confirmatory evidence has been found and Darwin's doubts continue to undermine his own hypotheses. Without reproducible data to fill in these gaps of knowledge, Darwin's whole thesis collapses. By definition,[496] *a hypothesis is either a suggested explanation for an observable phenomenon, or a reasoned prediction of a possible causal correlation among multiple phenomena. In science, a theory is a tested, well-substantiated, unifying explanation for a set of verified, proven factors. A theory is always backed by evidence; a hypothesis is only a suggested possible outcome, and is testable and falsifiable.*

After more than 150 years since Darwin's Origin of the Species, where are all the innumerable transition fossils that demonstrate Darwin's nearly imperceptible gradation to new kinds? How do we explain complexity as in the eye without precursor gradations in the Cambrian explosion of life? Why have the number of phyla and body plans decreased[497] rather than increased since the Cambrian?[498] Why isn't there a chemical mechanistic basis for the evolution of species? If as many conclude, Darwin's research findings (e.g., finches) apply in a rudimentary way to microevolution, BUT NOT to the macroevolution as Darwin claims, why isn't the "theory" being dismissed?

Today few scientists dispute microevolution. But macroevolution is an entirely different story. Macroevolution calls upon exceedingly long periods of time. *But the devil is in the details.*

---

496 www.diffen.com/difference/Hypothesis_vs_Theory
497 Ward, Peter D. and Donald Brownlee. 2007. Rare Earth: Why Complex Life is Uncommon in the Universe. Springer. Page 142.
498 Lewin, R. (1988). A lopsided look at Evolution: An analysis of the fossil record reveals unexpected patterns in the major evolutionary innovations, patterns that presumably reflect the operation of different mechanisms. Science, Volume 241, p. 291.

At a deeper, fundamental level of scientific scrutiny,[499] *evidence from physical and organic chemistry* should be brought forward describing the macroevolutionary transition by natural selection from one kind to another, if the scientific community is to proclaim that evolution is valid. Without the following kinds of mechanistic and/or phenomenological evidence, macroevolution is at best conceptual:

1) **Scientists need to produce the mechanistic and phenomenological processes for the origin of the vital compounds of life like the proteins that comprise DNA, and RNA, and how they have evolved with time**. Professor James M. Tour at Rice University concludes that, *although most scientists leave few stones unturned in their quest to discern mechanisms before wholeheartedly accepting them, when it comes to the often gross extrapolations between observations and conclusions on macroevolution, scientists, it seems to me, permit unhealthy leeway... I simply do not understand, chemically, how macroevolution could have happened... I will tell you as a scientist and a synthetic chemist: if anybody should be able to understand evolution, it is me, because I make molecules for a living.*

2) **Scientists need to produce the mechanistic and phenomenological chemical processes for the origin of a living, self-replicating cell in a prebiotic world.** DNA and RNA carry the instructions for making proteins, and proteins extract and copy those instructions as DNA or RNA... For decades, the favored candidate has been RNA... But RNA is also incredibly complex and sensitive, and some experts are skeptical that it could have arisen spontaneously under the harsh conditions of the prebiotic world.

3) **Scientists need to establish the mechanisms that led to self-replication (sex) by male and female animals.** According to Sinai, et al., a major objective of research on the origin of life is to find plausible self-replicating chemical systems...Formidable

499 Luskin, Casey. 2014. The Top Ten Scientific Problems with Biological and Chemical Evolution. Volume: More than a Myth. Chartwell Press.

difficulties confront the development of this narrative into a complete and rigorous theory of life's origin.

4) **Scientists need to establish the limitations of evolution in the face of sterility due to hybridization.** According to Dr. Norman Johnson, hybrids between closely related species are often inviable or, if they live, they're sterile. This hybrid unviability and sterility, collectively known as hybrid incompatibility, can reduce the exchange of genetic variants between species. Thus, hybrid incompatibility can be important in the process of speciation by acting as a reproductive isolating barrier (Coyne & Orr, 2004).

5) **Scientists need to establish the mechanistic processes that support the evolution of complex features like the eye and other organelles/organs.** How is it that the complex, compound, fully-faceted eye of the trilobite appeared in the Cambrian without predecessors? Charles Darwin himself acknowledged that it might seem absurd to think the eye formed by natural selection.

6) **Scientists need to develop and verify the chemical reactions that would lead to the birth of a living cell.** How is the guidance system for the cell encoded to deal with the routine and non-routine environmental stimuli?

7) **Darwin's concern for the missing fossil record has, if anything become a bigger issue than it was in Darwin's day**. How would Darwin answer his own concern if he had today's information?... why, if species have descended from other species by insensibly fine gradations, do we not everywhere see innumerable transitional forms?

8) *March 1, 2018 $5 Million Tech Prize Seeks Answer to Origin of Life:* In August of 2017, a group of entrepreneurs announced the Evolution 2.0 Prize of $5 million at Arizona State University.[500] The objective is said to encourage a breakthrough in understanding DNA coding that would have incredible impacts in artificial intelligence, cancer, genomics, and agriculture to name a few. According to Marshall, *"A good scientific experiment will show how genetic code somehow emerged spontaneously over*

500 Esson, Alex. $5 Million Tech Prize Seeks Answer to Origin of Life. Downloaded from frontlinegenomics.com June 2019. Also see www.herox.com/evolution2.0

millions of years." Scoffers believe that it can't be demonstrated. I think the prize was established to openly make a mockery of evolution as a theory! What do you think?

In conclusion, the "Theory of Evolution," from a scientific perspective, is not a theory at all but rather a concept. It's very foundations have never been illustrated or proven at the depth required by modern science to establish it as a theory. Science cannot explain how the first cells arose from inert material or for that matter, how life suddenly appeared in the "*Cambrian.*" For it is in the "*Cambrian*" that we see animals with **brains, eyes**, and other sophisticated organelles, organs, and structures that suddenly appear without a demonstrated "*evolutionary pathway.*"

CONCLUSION 7: A Case for Intelligent Design

Many, including Newton and Keplar, have argued that there are numerous seemingly unrelated fine tuning factors that point to the Intelligent Design of our Universe. The physical constants of our Universe provide an environment suitable for life, by definition, after all, life exists. The following list provides several finely tuned factors that make life within our universe possible:[501]

- *If the gravitational force were slightly stronger, stars would be more massive than our sun... these stars would burn too rapidly and too inconsistently to maintain life supporting conditions on surrounding planets. If it were slightly weaker, stars would have less mass than the sun and there would be no heavy elements for building such planets or life.*
- *If the strong nuclear force were slightly weaker, hydrogen would be the only element produced, an essential element of life as we know it. If it was slightly stronger, then hydrogen would be rare and life essential elements would be insufficient.*

---

501 Hugh Ross. 1991. The Fingerprint of God, Recent Scientific Discoveries Reveal the Unmistakable Identity of the Creator. 2nd Edition. Promise Publishing Co. Orange, California.

- *If the weak nuclear force were slightly changed, either there wouldn't be enough helium to generate heavy elements in stars, or stars would burn out too quickly and supernova explosions couldn't scatter heavy elements creating habitable planets in the universe.*
- *If the rate of expansion of the universe were slightly slower the whole universe would have collapsed before any solar stars like the sun could have settled into a stable burning phase. If the universe were expanding slightly more rapidly, no galaxies would have formed. This constant is estimated at 1:10^55.*

Location of our Solar System within the Milky Way Galaxy:

Our Solar system is located in a stable orbit within our galaxy. Numerous authors provide lists of the fine tuning of our Solar System. *The following provides just a few of the finely tuned factors commonly referred to that make our Solar System habitable:*[502]

- *Our Solar System is located in a safe zone of diffuse gas and low radiation, hundreds of light years across*
- *The Earth's orbit is in the "goldilocks zone" where temperatures allow water to exist as a liquid, essential for life. Our neighbors closer to the Sun experience high temperatures not conducive to life; whereas planets beyond Mars are too cold for life.*
- *The Moon stabilizes the Earth's tilted axis which provides seasons allowing for diversity of life.*
- *The position of the Solar system in the Milky Way and the size and position of the moon enable excellent observation of the Universe.*
- *The Earth is shielded by larger planets like Jupiter, Saturn, and Neptune that protect it from asteroids.*

---

502 Hugh Ross. 1991. The Fingerprint of God, Recent Scientific Discoveries Reveal the Unmistakable Identity of the Creator. 2nd Edition. Promise Publishing Co. Orange, California.

Fine Tuning of Planet Earth:

Planet Earth is located in a stable orbit within our Solar System. *Some of the finely tuned factors commonly referred to that make our Planet habitable:*[503]

- *Earth has abundant water, essential to life.*
- *Earth has a magnetic field and an ozone layer that protect life from harmful radiation. Lightning replenishes the protective ozone layer.*
- *Earth's atmosphere is ideal for life allowing sufficient light for photosynthesis. If oxygen levels were higher, fires would be more common. Yet Earth's plant life removes carbon dioxide and provides oxygen for higher life forms. Oceans absorb and distribute heat around the globe, and they act as a planet-sized CO_2 scrubber, and their high heat absorbing capacity prevents a runaway greenhouse effect.*
- *Earth's gravity slows the loss of water to space.*
- *Earth's plate tectonics recycle nutrients like a conveyor belt from the surface to the lower crust and mantle.*

From a statistical perspective, the existence of intelligent life forms arising by chance is not possible, yet here we are. According to astrophysicist Fred Hoyle:

"A common sense interpretation of the facts suggests that a super intellect has monkeyed with physics, as well as with chemistry and biology, and that there are no blind forces worth speaking about in nature. The numbers one calculates from the facts seem to me so overwhelming as to put this conclusion almost beyond question."[504]

---

503 Hugh Ross. 1991. The Fingerprint of God, Recent Scientific Discoveries Reveal the Unmistakable Identity of the Creator. 2nd Edition. Promise Publishing Co. Orange, California.
504 Hoyle, F. 1982. The Universe: Past and Present Reflections. *Annual Review of Astronomy and Astrophysics*: 20:16.

Nature undeniably provides evidence that the Earth's placement in in the Universe is ideally suited for organic life.

CONCLUSION 8: The Rocks Cry Out: The Younger Dryas and the Final Warning

Nearby planets and moons tell us that catastrophic events punctuate and dictate their fate. Mars and Venus have suffered visible catastrophic damage. Why should Earth escape anything less? The shattering of our planet and the rapid rotation of its continental plates gives testimony to a catastrophic past. Forty percent of the world's population lives within 100 kilometers of the coast. And according to the National Oceanographic and Atmospheric Administration (NOAA):[505]

- *The rate of sea level rise is accelerating: it has more than doubled from 0.06 inches (1.4 millimeters) per year throughout most of the twentieth century to 0.14 inches (3.6 millimeters) per year from 2006–2015.*
- *In many locations along the U.S. coastline, high-tide flooding is now 300% to more than 900% more frequent than it was 50 years ago.*

The pattern of sea level rise is not new. Oceans have been steadily rising since the end of the Younger Dryas. The 7th cycle is underway and that brings with it another invasion of the seas. Greenland alone has enough ice to raise global sea levels by 23 feet and if the ice sheets of the Antarctica were to completely melt, the oceans would deepen by more than 200 feet.[506]

The Laurentide ice sheet that stretched as far south as New York State and Ohio some 20,000 years ago had retreated to eastern Canada, just across the water from Greenland, by

505 Lindsey, R. 2019. Climate Change: Global Sea Level. NOAA. Climate.gov.
506 Biello, D. 2008. A Deep Thaw: How Much Will Vanishing Glaciers Raise Sea Levels? Scientific American. Scientificamerican.com

roughly 11,000 years ago thanks to increased sunlight (due to the periodic wobble in Earth's axis known as precession). It then completely disappeared by 6,800 years ago in two geologically rapid bursts, shedding enough ice to raise sea levels by as much as four feet (1.3 meters) per century.[507]

The current sea floor represents about 70% of the surface of the Earth's crust. How is it that the *today's entire sea floor* was laid down in the waning moments of Earth's history, *considered by geologists to represent less than 4% of the total elapsed time* since Creation. *How is it that the vast majority of this same sea floor was formed in just the last 1.5% of geologic time?* Where did the replaced ocean crust go? We know that part went into the building of mountains, and part was subducted. Could all of this have taken place at 2 centimeters a year? How could the Farallon Plate, over a thousand miles of subducted ocean crust, still be intact miles beneath the continental United States? How could it maintain its strength and identity for nearly 65 million years without being totally assimilated by the heat of the mantle?

Two enormous outbursts of carbon dioxide are recorded in the rock record from the beginning of the Cambrian to the Present. Both appear to be related to accelerated periods of plate tectonics and associated volcanic activity. Carbon dioxide during periods of rapid plate tectonics and vast outpouring of magma were 10 to upwards of 30 times present levels. Could these have been driven by some combination of Core-Mantle nuclear events and/or extraterrestrial impacts? *Global warming is driven by nature; far, far less by man. Global warming has been greatly sensationalized. Global warming is just the beginning of the 7th Cycle: APOCALYPSE NOW. Afterall, sea levels have been steadily rising for the past 10,000 years.*

It seems a horde of volcanoes began spewing just before the Cambrian, and their activity reached a peak during the Dead Interval. "We hypothesise that CO_2 outgassing from continental volcanic arcs drove major climate shifts," says

---

507 Biello, D. 2008. IBID.

McKenzie.[508] **During the maximum advances of the ice sheets the entire continental shelves were exposed, including the Bearing land bridge between North America and Asia. Today, we are in an interglacial period. Sea level has been rising slowly during the past 10,000 years...[509]**

CONCLUSION 9: Feasibility of Cyclical Core-Mantle Boundary Nuclear Explosions

Natural Nuclear Reactors were predicted by Kuroda and one such reactor has been confirmed at Oklo, Gabon. A number of scientists are once again proposing that natural nuclear reactors might form at the core-mantle boundary. **The vast majority of the heat in Earth's interior—up to 90 percent—is fueled by the decay of radioactive isotopes like Potassium 40, Uranium 238, 235, and Thorium 232 contained within the mantle. Could the recycling of ocean crust provide feed for these reactors? Could these form by a cyclical process of: core-mantle concentration of constituents, generation of heat, explosion, scattering of constituents, followed by repeated concentration of constituents at the Core-Mantle boundary:**

> *...a shock wave created by rapidly expanding plasma resulting from the explosion disrupts and expels overlying mantle and crust material.[510] Thus Earth's Fe-rich alloy core, with limited U and Th, is a probable site. Here we postulate that the generation of heat is the result of three-body nuclear fusion of deuterons confined in hexagonal FeDx core-centre crystals; the reaction rate is enhanced by the combined attraction effects of high-pressure (~364 GPa) and high-temperature (~5700 K)... The possible heat generation rate can be calculated as 8.12×10^{12} J/m³, based*

508 Brahic, C. 2014. Volcanic mayhem drove major burst of evolution. Newscientist.com

509 Columbia University. downloaded 2019. Cenozoic Climate: From the Greenhouse to the Icehouse. Columbia.edu.

510 de Meijer, R.J. and others. 2013. Forming the Moon from terrestrial silicate-rich material. Chemical Geology, Volume 345. Pages 40-49.

on the assumption that Earth's primitive heat supply has already been exhausted... AND Meijer and van Westrenen reported nuclear fission of U and Th as heat generation sources at the mantle boundary within Earth's core, based on the distribution of an isotope of Nd in rocks. Bao noted that there are many heat producing elements (U and Th) in a calcium perovskite reservoir at the base of the mantle.[511]

CONCLUSION 10: Time- a Matter of Confusion

The Scriptures tell us that Satan deceives and confuses mankind. Like evolution, the debate over the Age of the Universe has caused confusion and driven many from the Church and a belief in God. If the age of the Earth is on the order of 6000 years, a great number of catastrophic events, since the beginning of the Cambrian, would need to be compressed into the short time interval referred to as the year-long Biblical Flood. Is this possible? **With God all things are possible.[512]** Alternatively, there is another logical argument that supports the Creation days recorded in Genesis. It is based on the perspective that God is outside of our time-space Universe, whereas, time viewed by mankind is confined within time-space.[513] These lines of reasoning are neither fact nor theory; they are hypotheses. The age of the Universe is unessential to salvation. Scientists need to lay aside unresolved decisive differences.

Scientists within the Christian Community need to unite to provide the lay community with the evidence that GOD EXISTS. Fortunately, both groups of scientists have a foundation to build upon. Most agree that God is eternal and not confined to our natural world. We also agree that life is created by Intelligent Design, not by evolution. **...choose you this day whom ye will serve...[514]**

---

511 Fukuhara, Mikio. 2016. Possible generation of heat from nuclear fusion in Earth's inner core. Scientific Reports. Volume 6. Article number: 37740.
512 Mark 10:27.
513 Ross, Hugh. 2004. IBID; Schroeder, G. 2013. IBID.
514 Joshua 24:15.

POSTSCRIPT

In the near future, I will be releasing a series of books that I believe will add to the compelling basis for the existence of God. The cornerstone is:

Guardians: Keepers of God's Secret Code.

ABOUT THE AUTHOR

Dr. Don Alexander served as lead geochemist in US NRC's Office of Nuclear Materials Safeguards and Security (NMSS) where he helped develop regulations for the disposition of High Level Waste. He later worked for DOE's Office of Civilian Radioactive Waste Management (OCRWM), serving as a liaison to the Nuclear Energy Agency in Paris. At OCRWM, he managed the development of Site Characterization Plans for the evaluation of a U.S. National Repository. With DOE's Office of Energy Waste Management (EM), he led International Technology Program missions to the Former Soviet Union (FSU), Japan, France, England, Italy, Germany, Hungary, and throughout Europe. He led a mission to the FSU and established the first exchange of scientists working on nuclear waste programs between the United States Department of Energy's National Laboratories and the Former Soviet Union. In that capacity, he negotiated for the transfer of Russian data on the effects of radiation on workers and civilians in the Urals. Don taught for a number of years as a Clinical Professor in the Department of Environmental Health Sciences, Tulane University School of Public Health and Tropical Medicine and served as PhD Co-Chair for three dissertations.

Don holds a Doctorate of Philosophy in Geology, specializing in Geochemistry, from the University of Michigan.

Back Cover Credit: Adopted from image by Elliot Lim and Jesse Varner, 2008. *Age, spreading rates, and symmetry of the world's ocean crust.* Cooperative Institute for Research in Environmental Sciences, University of Colorado & NOAA National Geophysical Data Center. Data Sources: Muller, R.D., M. Sdrolias, C. Gaina, and W.R. Roest. Public Domain. Interpretations are solely those of the author.

The Rocks Cry Out, warning of an ominous unstoppable disaster already set in motion.

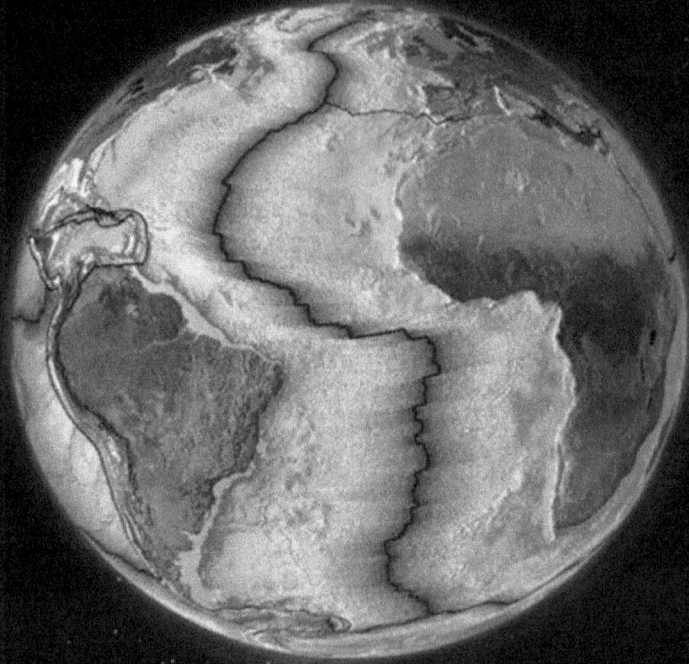

How could entire new ocean floors be so rapidly formed in just the last 2% of geologic time? Evidence carved in stone by an extinct civilization and found in the Rock Record provides undeniable evidence of Global Flooding, Fountains of the Great Deep, Catastrophism not Uniformitarianism, Challenges to Evolution, Seismic Wave induced Plate Tectonics, a Case for Intelligent Design, and a warning that the 7th Apocalypse Now is underway.

www.ingramcontent.com/pod-product-compliance
Lightning Source LLC
Chambersburg PA
CBHW031248090426
42742CB00007B/356